THE GREGYNOG PRESS

A HISTORY OF THE
GREGYNOG PRESS

DOROTHY A. HARROP

PRIVATE LIBRARIES ASSOCIATION

MCMLXXX

Published by the Private Libraries Association
Ravelston, South View Road, Pinner
Middlesex, England

Printed and bound in Great Britain by
T. & A. Constable Ltd, Edinburgh

Designed by David Chambers

2,500 copies, of which 750 are for sale
SBN 900002 63 8
100 copies signed by the author
SBN 900002 73 5

TO THE MEMORY OF

DORA HERBERT JONES

1890-1974

Contents

Illustrations

PLATES

Preface

IT IS FASHIONABLE IN WRITING ABOUT THE WORKS of private presses to-day, when most items, good, bad or indifferent, command an unnaturally inflated price, either to exalt them to the point of idolatry or to decry them as specious objects designed to be admired but not read. Custom would seem to demand a formulation of reasons for their creation, a seeking beyond the books themselves for some assessment of their contribution to the evolution of the printed book of our time. Writers are at pains to find analogies, seminal changes, new directions and approaches ascertainable in the works examined and, beyond that, to discern the philosophy of their creators. *What is it all for?* What motivation lies behind the expenditure of so much time, energy, money and thought? Such searchings may be perfectly valid and often make interesting reading, yet surely this is to miss the real *raison d'être?*

In common with other creative works most private press books are conceived in love and born, by trial and error, through hours of commingled joy, excitement, exaltation, desperation, disappointment and despair. The press is not *for* anything other than the satisfaction of the creative urge in its owner. This is not to agree entirely with one writer that

> 'the joy of a private press is indifference to public demand'[1]

for most private printers delight in public approbation of their work and are thrown into the depths by castigation or, worse, by neglect. *Aims* are frequently formulated by private printers but usually from hindsight when the course is already embarked upon and when other people begin to ask them why they do what they do.

Much time has been and is spent in the formulation of all-embracing definitions of the term 'private press', but many of the attempts would seem to contain but half truths. The only definition which can undeniably be said to be true of them all is that a private press is one where the owner calls the tune without reference to outside let or hindrance. To break even in financial terms or, better still, to make a profit, which private presses have rarely done, may be part of the vision but is not the prime consideration. The owner's satisfaction, pleasure and education come first and are paramount in the creative endeavour. Private press books are seldom items of unparalleled technical perfection but often flawed works which are of interest as much for what they reveal concerning the personality of their creator as for their aesthetic attributes. For this reason they must lie forever outside the wider bounds of general printing history. In the pursuit of perfection one is reminded that the goal is entirely subjective. I have in my files a letter from a commercial publisher concerning the appearance of a rival's products, commonly held to be of an almost irreproachable standard, which contains the sentence, 'I should hate my books to look like those of X which are too perfect to handle, let alone read'.

Though the days of the great British private presses are over, stilled forever by the change in our world which occurred with the advent of the Second World War, interest in their works remains not only unabated but growing, due largely to the increased number of book collectors among ordinary folk and to the growth of investment collecting. Thus it seems surprising that there have been only two recent attempts at writing full scale histories of individual private presses in the British Isles. I refer to *The Stanbrook Abbey Press*, written and published by the Community in 1970 and Liam Miller's *The Dun Emer Press, Later the Cuala Press* published by the author's own Dolmen Press in 1973.

Now, while its memory is still fresh in the minds of people who visited it in its heyday, and while there are still alive some who contributed significantly to the making of its books, which may still be acquired by the ordinary collector, though at ever increasing prices, I have chosen to set down the annals of the Gregynog Press.

There are normally individual or unusual features which serve to distinguish the activities of one private press from another, and the Gregynog Press is no exception. The two most outstanding features in this case are the lack of knowledge or skill on the part of the owners and the fact that all aspects of the books' design and making were carried out under one roof. These alone render the press virtually unique among British private presses of the twentieth century. Books from the Doves, Ashendene, Vale, Golden Cockerel and other presses bear unmistakable marks of ownership, yet no Gregynog book owes anything in its appearance, unless it be a love of illustrated and decorated books in general, to the Davies's who owned the press. Neither of the two sisters had any knowledge of typography or printing. Neither ever set a line of type or pulled a sheet from the press. Gwendoline Davies visited the press but rarely, and then usually to escort visitors, Margaret Davies hardly ever. They knew what they liked and gave approval or otherwise to a staff which kept its distance, to carry out the work in their name. Their role lay in providing substantial sums of money to support the press, in providing accommodation and machinery and, through their intermediary, Dr. Thomas Jones, advice from outside literary and technical experts.

The Davies's were extremely wealthy, and no press in this century, with the exception of Count Kessler's Cranach Presse in Weimar, has had such munificence lavished upon it. No expense was spared in terms of time spent or quality of materials to make a book as perfect as they knew how. If setting or presswork did not come up to standard, work began again. Several books were laid aside uncompleted, and, in one instance, a book was completed but never issued because it fell short of the Davies's expectations. It is inconceivable that the means for such care and concern will ever be available again.

The unity in conception of design and execution in Gregynog books has frequently been noted by writers on the subject, yet this is scarcely so remarkable when one remembers that design layout, composition, presswork, design and execution of woodblocks, hand-colouring and the complete binding of the books all took place in a handful of small rooms, on two floors set one above the other. Continuity over a period of eighteen years and four different controllers was provided not only by the consistent attitudes of a small permanent press board but also by the continuing service of a small number of dedicated staff who served through more than one reign of control.

Though not unique in this respect, the press was unusual in that a great proportion of its books were illustrated works. All but nine of the forty-two published have illustrations, and most of the remainder display decorative initial letters. This aspect derives no doubt from the fact that the Davies's main interests were artistic and not literary.

One feature does render Gregynog unique among the major private presses. Not only was the press situated in Wales, but seven of its books were actually printed in the Welsh language; one work is bilingual and eleven others, though printed in English, are by Welsh authors or have

Welsh connections. The first fourteen books, published during the first seven years of the press's existence, fall into one or other of these categories.

In writing the history I have chosen to chronicle events as they occurred and as they affected either the Davies's, their advisers or those whom they employed. Too often our knowledge concerning the workings of a private press is based on a bare outline of facts, bulked out with critical appraisal of its printed works. Here I have attempted to recreate the atmosphere of an epoch, leisured and cushioned by wealth, and of a place which still remains but with a much changed function, and to evoke something of the triumphs and disappointments of those who worked in the press.

Appended are a revised bibliography of the forty-five books printed at Gregynog which amplifies the information and corrects a few errors found in the bibliography of Doctor Thomas Jones's *The Gregynog Press: A Paper Read to the Double Crown Club on 7 April, 1954* and also the first published checklist of ephemera. Much of the ephemera is scattered in private collections, and I am grateful to the various owners for having allowed me access to the material in their possession. While I cannot claim that the list is in any way definitive, it represents the fruits of five years' search and must cover the great majority of such items which emanated from the press.

I owe a debt of gratitude to many people who have assisted in the compilation of the history. I am particularly grateful to the following: Baroness White of Rhymney and Mr. Tristan Jones for permission to consult the papers of their father, Dr. Thomas Jones, C.H.; the late Miss Jessie Adams; Mr. John E. Bowen; Mr. Sydney Cockerell; the Hon. Islwyn Davies; Mrs. Gwen Edwards; the late Dr. Joan Evans; the late Mr. George Fisher; Dr. Loyd Haberly; Miss Gertrude Hermes; the late Mr. Herbert John Hodgson; Mr. Blair Hughes-Stanton; Dr. Glyn Tegai Hughes; Mr. Idwal Jones; Mrs. M. McCance; and the staff of the National Library of Wales. Thanks are also due to those who lent photographs, including some of those just listed, as well as Mrs. P. Bray, Mrs. P. Bunting, Mr. Bert Hodgson, the National Museum of Wales, Mrs. E. Wardrop, and to the University of Wales Press for the loan of a photograph and for allowing the reproduction of pages from the books. My special thanks must go to Mr. Hugh Herbert-Jones for allowing his books to be photographed, and to Mr. Colin Franklin, Mr. Idris Jones and the late Mr. James Dearden who kindly read the text and made helpful suggestions, to my colleague, Mr. John Harris, for assistance with the bibliographies and to Mr. David Chambers for much expert advice. My debt to the late Mrs. Dora Herbert Jones is incalculable. Her complete collection of Gregynog books and extensive collection of ephemera were ever at my disposal until her death. Her wide knowledge of the Press, which was tinged with a keen critical awareness of its strengths and weaknesses, was most generously shared over several years. Her interest and encouragement in the project throughout were such that whatever merits it may have derive mainly from that source. In small token of repayment for much kindness I dedicate this book to her memory. My deep regret is that she did not live to see its completion.

D.A.H.

Llanbrynmair, 1975

NOTES

1. Colin Franklin. *The Private Presses*. Studio Vista. 1969.

Engraving by Robert Maynard from Penillion Omar Khayyâm.

I

The Davies Family

IN THE EARLY PART OF THIS CENTURY THE Davies family was still comparatively nouveau riche. Its fortunes were amassed by David Davies who was born in 1818, of humble parentage, at Llandinam, a small Montgomeryshire village nestling at the foot of the southern slopes of a range of hills which fall steeply into the upper Severn valley. 'Top Sawyer', as he is still referred to in Wales from his original occupation which was timber sawing on a contract basis, was destined to become the archetypal self-made capitalist in an age of industrial expansion. In 1851 he married Margaret Jones of Llanfair Caereinion in the same county. From his first civil contract, which was to lay the foundation approaches for a new bridge over the Severn in his native village in 1846, he rose to become a major partner in the construction of seven railways between 1855 and 1871. By 1864 Davies, now a man of substance, had built Broneirion, the house on the southern banks of the Severn, just outside Llandinam, in which he lived until his death.

Beginning in 1864 David Davies extended his industrial interests to the coalfields of the Rhondda, and by 1855 had sunk no less than six pits. This led to the formation of a limited liability company, The Ocean Coal Company Limited, in 1887. By this time the Cardiff docks had become inadequate to handle the company's output which led Davies to promote the Barry Dock, together with a railroad connection from the Rhondda mines, which was eventually completed in 1889.

Increasing wealth enabled him in 1884 to purchase the substantial Llandinam estate together with Plâs Dinam which became his son's home two years later. From the 1870s Davies played a significant role in public life. From 1874 to 1886 he served as Liberal Member of Parliament for the Cardigan district. He was one of the original governors and treasurer of the newly founded University College of Wales at Aberystwyth and in 1889 was elected to represent Llandinam on the first Montgomeryshire County Council. A Calvanistic Methodist, Davies was rigidly opposed to drink and puritanical in his observance of the Sabbath. A certain homespun quality and shrewdness of character are evident in the fine portrait of him which Ford Madox Brown painted in 1874.[1] He never forgot his humble beginnings; indeed he was proud of them and was always anxious to help any truly deserving cause, be it that of an individual or an institution. His liberality was to earn him enduring respect and affection among Welsh people. Considerable sums of money were distributed to various religious and educational causes in the Principality throughout his later years, a practice continued by his descendants.

When David Davies died in 1890, his only son, Edward, succeeded to the family estate and fortunes. Edward, on whom so much responsibility weighed heavily, enjoyed indifferent health. Eventually his mind became affected and he died at the early age of forty-seven in 1898.[2] An extensive block of chemistry laboratories was given in his memory by his widow and children to University College Aberystwyth in 1907. He left a son, David, born in 1880 and two daughters, Gwendoline Elizabeth, known simply as Gwen, born in 1882 and Margaret Sidney, always known as Daisy, born two years later.

David, who was created first Baron Davies in 1932, can justifiably lay claim to being the greatest Welsh public benefactor of his age. In 1911 he founded, with an initial grant of £150,000, the Edward VII Welsh National Memorial Association to combat tuberculosis, a cause for which he later provided an administrative centre, together with one for the Welsh League of Nations Union, in the Temple of Peace and Health opened in Cardiff in 1938. He remained president of the association until his death in 1944. He also endowed research centres in Cardiff and in 1921 the David Davies Chair of Tuberculosis at the Welsh National School of Medicine. Other significant benefactions included large sums of money to the Presbyterian Theological College at Aberystwyth, to the National Library of Wales of which he became president in 1927 and to the University College at Aberystwyth for the foundation of chairs of International Politics and Colonial History. He served as a Liberal Member of Parliament for Montgomeryshire, 1906–1929 and was Parliamentary Secretary to Lloyd George, 1916–1917. He was a member of the Council of the University Colleges of Wales and of the University Court and, from 1926, President of the Council of University College, Aberystwyth. Lord Davies's chief interest lay in the promotion of world peace which led him to become founder and trustee of the League of Nations Union and permanent secretary of its Welsh Council. His ideals for peaceful co-existence were relentlessly pursued in speech and writing throughout his life.

His two sisters, who also inherited great wealth, were no less generous. Their chief beneficiaries were University College Aberystwyth, where they established the Gregynog Chair of Music, the National Library of Wales, the National Museum of Wales and, ultimately, the University of Wales to which Margaret Davies donated Gregynog. Unlike their brother, who had attended Merchiston Castle, Edinburgh, and King's College, Cambridge, they were educated as was thought fitting for ladies of their time at private schools and never attended a university. Following family tradition, they were brought up as Calvanistic Methodists. Though born in Wales, of Welsh stock, and staunch lifelong supporters of various Welsh causes, they were not markedly Welsh in temperament or outlook and had scant knowledge of their native tongue. Their literary and artistic inspiration stemmed mainly from outside the Principality. They were intensely interested in music, in art and in crafts of many kinds, though lacking any formal training, were not conspicuously knowledgeable concerning them. In this respect they were dilettante, but such artistic causes as they espoused they supported with enthusiasm.

Gwen, the elder, more sensitive and more genuinely artistic of the two sisters, was in her early days an accomplished violinist and owned a Stradivarius. When her fingers were rendered less sensitively supple by the blood disease which shadowed her life, she took to playing the organ. Daisy Davies played the harp and sang. She briefly attended the Slade School in her youth as an external student and later was tutored privately in painting, becoming a reasonably talented amateur painter of landscape. Both sisters enjoyed travel. Gwen visited the United States shortly after leaving school, and later on they frequently indulged in excursions abroad, visiting such exotic places as Petra and Damascus, in between visits to France and other western European countries.

Their youth was spent in the family home at Llandinam under the watchful eye of their stepmother who was also their aunt, their father having married the sister of his first wife who died in 1888. Their erstwhile governess and lifelong friend who later became their stepmother's companion was Jane Blaker, the sister of Hugh Blaker,[3] a diversely talented artist, connoisseur, art critic and dealer and one time curator of the Holburne of Menstrie Museum in Bath. Blaker was one of the first people in Britain to appreciate the work of the French Impressionist and

Post-Impressionist painters, and it was under his guidance that the Davies sisters, beginning in 1908, amassed the outstanding collection of paintings which now grace the National Museum of Wales in Cardiff.

Neither of the sisters would have laid serious claim to being an art connoisseur, having inherited no tradition of art appreciation and having undergone no formal education in art, but from youth they were keen exhibition and gallery visitors. Acting on Blaker's advice, they bought paintings consistently over a number of years, mainly through agents in London and Paris, apart from a few which Gwen Davies bought personally in Paris during the First World War. By 1924 their collection numbered some two hundred items, nearly half of them French nineteenth-century paintings, the remainder being made up of a few old masters, a number of conventional nineteenth-century continental works and some English paintings and drawings spanning the late eighteenth to early twentieth centuries. In addition there were some pieces of sculpture including works by Rodin and Degas. As far as late nineteenth-century French paintings went it was a pioneer collection in Britain and an astonishingly rich collection by any standards. After the sisters had embarked on their Gregynog venture the purchase of paintings ceased, and no more were acquired until Daisy Davies started some desultory collecting after 1934. The French paintings which early on aroused much interest, and even antagonism, were frequently lent for exhibition purposes. After 1924 the art collection was housed at Gregynog until it was eventually bequeathed to the National Museum of Wales.

In pursuit of artistic interests, enlivened by a degree of social life and travel, and on the part of Gwen Davies by horse riding,[4] the sisters spent their days until the advent of the First World War. In July 1920, they acquired the Gregynog estate, some eleven miles north-west of Llandinam, which by then consisted of the house, seven hundred and fifty acres of park and farm land and a number of farms.

NOTES

1. The painting is reproduced in John Ingamells' *The Davies Collection of French Art*. National Museum of Wales. 1967.
2. 'If you want to know what [my father] was like, think of Daisy – honest, kind, unselfish, straight as a die and as unbending . . .' Letter from Gwen Davies to Dr. Thomas Jones, dated 26 March 1929.
3. Hugh Oswald Blaker, 1883–1936.
4. 'It was only on a horse that I could feel myself and "master of my soul" in those days and the only time one was free from the vigilant eye of dear Mother . . .' Letter from Gwen Davies to Dr. Thomas Jones, dated 26 January 1943.

2

Gregynog

GREGYNOG HALL,[1] SECRET AND REMOTE, LIES IN a shallow, wooded valley, high up in the green undulating hills which lie to the north-west of Newtown. The earliest mention of this ancient seat is found in a twelfth-century poem by Cynddelw Brydydd Mawr. Its early history is obscure, but from the fifteenth century it was the home of the Blayney family, traditionally descended from Brochwel Ysgythrog, a sixth-century Prince of Powys. In what manner the early owners lived is not known, but in the present house hangs a water colour by Moses Griffith[2] depicting an unpretentious late sixteenth- or early seventeenth-century house, of two storeys, built in red brick and roofed with blue slates. This clearly is the house which preceded the present one. The carved wooden panelling, removed from the earlier dwelling, which is now in the room known as the Blayney Room is dated 1636, so that the house in which Charles I rested on Sunday, 21 September 1645, three days before the defeat at Rowton Moor, was probably relatively new.

The Blayneys, a conspicuously uninteresting family, continued to live at Gregynog until the death of Arthur Blayney in 1795 when the estate was bequeathed to an English cousin, Susannah Weaver, who had married Viscount Tracy of Toddington in the north Cotswolds. Three years later their heiress married her cousin, Charles Hanbury, who thereupon enlarged his name to Hanbury-Tracy and who is remembered as the architect of the great neo-gothic house at Toddington, built between 1820 and 1840. On the occasion of Queen Victoria's coronation, Hanbury-Tracy became the first Lord Sudeley. He was responsible for purchasing additional land at Gregynog, so that by 1820 the estate comprised no less than 18,000 acres. Sudeley's third son, Henry Hanbury-Tracy, lived there for a number of years but, finding the old house cramped and inconvenient, persuaded his father to enlarge it. The virtual rebuilding, completed in 1837, rendered the house unrecognisable. Only the carved panelling, already referred to, remains a visible link with the former building.

The new house was again built of red brick but rendered and painted to imitate the old black and white half-timbered style of which much that is genuine may still be seen in Montgomeryshire.[3] The general plan of the centre portion of the former house was retained, but the rooms were given extra height, an additional storey was added and a new wing built on. Formerly the house had fronted on to an area of marshland which was now drained to make a sunken garden, and a new main driveway was constructed.

The Sudeley fortunes were sadly diminished under the fourth and fifth lords, largely due to the failure of commercial and industrial enterprises in which they engaged, most notably the Newtown woollen flannel industry. In 1895 the estate was sold to Lord Joicey who retained it until 1913 when it was split up, most of the 235 farms and much of the land being sold to tenant farmers. At this point the hall and remaining 750 acres were let to Quintin Dick, Esq., a well known dog breeder. In 1920 it was bought by Gwen and Daisy Davies.

The house and grounds outwardly remain, under its present owners, the University of Wales, much as they were in the days when the Davies's bought the property except that, under the sisters, the gardens were developed and improved, and the University has infilled the back of the E-shaped house with a modern brick extension in order to increase accommodation. It is a peaceful setting. Approaching it up any one of the three tree-lined drives which meander through the park, one comes suddenly and unexpectedly upon the black and white frontage hitherto tucked away out of sight. Gregynog is best seen in winter, its lawns covered in snow and its massive trees glinting with hoar frost, or on a still June day when the house is encircled with a blaze of brilliant colour from banks of rhododendron and azalea. Sham the house may be, but its solid mass, rising from sunken lawns, in a sylvan setting, beautiful and tranquil, has warmed the heart of many an architectural purist.

NOTES

1. The origin of the name is obscure, but it is possibly derived either from *grugin* – the old form of the Welsh word for heather or from *Grugyn* – a name appearing in the *Mabinogion*. *Og* is added as a place name ending.
2. Moses Griffith, 1747–1819, water-colour painter and illustrator of Pennant's *Tours*.
3. Montgomeryshire was absorbed by the new administrative county of Powys in April, 1974.

3
The Grand Idea

DURING THE GREAT WAR, FROM 1916, GWEN DAVIES provided and ran a canteen for French troops near Troyes. She was assisted by Daisy, who later moved to run a similar enterprise near Rouen, and by a friend, Dora Herbert Jones, who features later in the Gregynog story. After the war, the two sisters decided to purchase Gregynog. The precise motivation is not known. According to one source, the reason, which I have been unable to substantiate, was that there was a bid to dismantle the house and ship it to the United States, and the Davies's wished to preserve it. Even allowing for the fact that many British houses have found their way across the Atlantic, such a removal seems unlikely due to its size and lack of architectural pedigree.

Following a public auction of furniture in May 1920, the property was legally conveyed to the sisters on 31 July. They had no intention of living in the house but spent their time devising schemes of an altruistic kind for its future use. The idea uppermost in their minds was that it should, in some way, benefit some of those whose lives had been torn asunder by the ravages of war. They envisaged a centre where ex-servicemen and others could be rehabilitated by receiving training in some kind of craft in beautiful and peaceful surroundings.

Behind these schemes there lay an element of appeasement. The Davies's wealth served as a constant reminder that it had been gained largely as a result of labour exploitation in the South Wales coalfields. Daisy was to expend much time and energy in an attempt to improve social welfare by the promulgation of girls' clubs, convalescent homes and similar ventures, while it was Gwen's lifelong ambition to bring colour and meaning into lives otherwise devoid of them. Their brand of philanthropy was to extend beyond the donation of money to various deserving causes. Gwen Davies in particular ardently desired to give something beyond mere cash and was prepared to spend her days trying to achieve her aim. Alas, her wealth created a barrier between her and those whom she longed to help, and as she grew older the cheques were written with an increasingly heavy heart. It proved a barrier too to meaningful communication with employees, to friendship often and, more importantly, to love, for the possession of great wealth so often breeds an innate terror of exploitation, and so it did with her.

From 1920 the sisters turned for help and counsel to Dr. Thomas Jones[1] who was currently Assistant Secretary, and later Deputy Secretary, to the Cabinet. They had become acquainted as a result of Jones's collaboration with their brother when, from 1909 to 1910, Jones had served as Secretary to the Edward VII Welsh National Memorial Association. T. J., as he was always affectionately known to friends and acquaintances, was to become the great ally in all their undertakings, acting as friend, confidant, adviser and their living link with the world outside Wales. No vital step was taken or decision made without recourse to his advice. In the absence of his constant help, support and encouragement it is doubtful whether any of their enterprises at Gregynog would have progressed beyond the blueprint stage.

That first summer T. J. gave up his holiday to stay in Tregynon, the village bordering the northern end of the Gregynog park, while he interviewed prospective domestic and garden staff. He also secured the semi-official services of Sidney Kyffin Greenslade, A.R.I.B.A., of Exeter, architect of the National Library of Wales,[2] to supervise building alterations at Gregynog. It is obvious from an early undated letter which Gwen Davies wrote to T. J. that she considered some of Greenslade's proposals over extravagant,

> 'He wanted to remove all the downstairs windows and put in casements with leaded lights; of course, it would have looked awfully nice, but it would cost £500.'

Greenslade acted for the sisters for a little over a year when he pleaded heavy commitments elsewhere and was replaced by Sir Edwin Hall,[3] a London architect noted for the design of several large hospitals and other public buildings.

While ideas were still fermenting concerning the uses to which Gregynog might be put, thought was devoted to obtaining a controller whose main task would be to organise instruction in several crafts. Once again the advice of Hugh Blaker was sought. T. J. was in favour of securing a Welshman for this key position, a notion which did not recommend itself to Blaker, particularly as he had a likely candidate in mind. His choice was Robert Ashwin Maynard, trained originally as an

Gwendoline Davies, 1937.

Margaret Davies, c. 1930.

Thomas Jones, 1936.

Gregynog Hall: the press room to the right, rear. △

John Hugh Jones, unknown, Elsbeth Herbert Jones,
Dora Herbert Jones, Hugh Herbert Jones, 1928.

architect at the London Polytechnic School of Architecture under Paul Waterhouse who had encouraged him to study painting. Prior to the Great War Blaker had been in touch with Maynard, then earning a successful living as a theatrical scene painter while attending evening classes conducted by Walter Sickert, Spencer Gore and Harold Gilman. His tutors were unanimous in considering him an artist of promise and tried to persuade him to quit scene painting, but Maynard was then a young man of twenty-five, already married, with no private resources, and prudence forbad reckless embarkation on so uncertain a career.

As early as 1913 Blaker had offered to seek the patronage of the Davies sisters on Maynard's behalf but had warned that they had previously received shabby treatment at the hands of one young artist whom they had supported which had rendered them cautious in such matters. He had also pointed out that they were slow in making decisions. In 1914 Maynard exhibited two paintings in the second exhibition of The London Group, held at the Goupil Gallery. He later recalled,

> 'I was tremendously proud, even to the extent of appearing in a ridiculous costume, surmounted by a topper with a flat brim which I had acquired in Paris. My first and last appearance on any stage as an exhibitionist!'[4]

The resourceful Blaker accompanied the Davies sisters to the exhibition and was careful to draw attention to Maynard's work with which they were much impressed, and it seems highly likely that they would have helped the artist financially had not the war intervened. Maynard served with the First Battalion, the London Scottish Regiment during the war and was subsequently commissioned in the Royal Fusiliers. His talent as a soldier apparently precluded his talent as an artist from being used in the services of camouflage. He fought in France, where he was wounded, and clearly hated it all. From this experience he emerged a disillusioned man and, after the war, accepted a post as salesman for a Shrewsbury manufacturer of cattle medicines! He was still in touch with Blaker who in 1919 mentioned the possibility of a post at Gregynog as organiser of crafts, but Maynard was not enthusiastic, protesting that he knew nothing whatever about crafts.

As a result of a chance meeting between him and Daisy Davies, accompanied by Jane Blaker, which took place on the platform at Machynlleth railway station early in 1920, Maynard was invited to visit Gregynog, a visit which was followed by several more, during the course of which plans for a craft centre were discussed, but no post was actually offered to him. The sisters obviously charmed him, for he later recalled,[5]

'[They] were very different from what I had expected. To my surprise, there was no side, no parade of wealth, in fact nothing to suggest that they owned the whole bag of tricks. They were simple and unaffected but just a trifle more earnest than I thought necessary. I liked them for it and quickly sensed the immense possibilities of the plans they were making.'

Fearing that his plan might yet come to nought, Blaker later engineered another meeting between Maynard and the sisters at the Victoria and Albert Museum, following which they repaired to the Davies's flat, 3 Buckingham Gate, for tea.

'I remember that the evidence of their wealth was more apparent at the flat than in the wilds of Montgomeryshire, and that it all went to my head!'[6]

It was finally agreed that Maynard should relinquish his post and spend some months in London, at the Davies's expense, learning as much as he could about the various crafts in which they had expressed an interest. Following this he was to go to Gregynog as Controller at a salary of £500 per annum, a sum less than that which he was already earning.

Maynard took up his appointment in February 1921 and spent the following eighteen months in London, at various establishments where crafts were taught, attempting to acquire the knowledge he lacked. Much of this time he spent receiving instruction in printing from J. H. Mason[7] at the Central School of Arts and Crafts and in acquiring the art of wood engraving in which he was largely self-taught. It soon became obvious that printing and wood engraving were the crafts to which he was most attracted.

Interest in printing sprang also from another quarter. The future of the Shakespeare Head Press was in jeopardy in 1921. The Press had been founded in 1904 by Arthur Henry Bullen as part of an attempt to form a national memorial to Shakespeare and a centre of English-speaking scholarship at Stratford-upon-Avon. Various bodies to which he had applied for support had refused funds, and Bullen had died a disappointed man in 1920. Individuals to whom he had turned, with some success, in the past had included the Davies sisters and their stepmother. A letter from Gwen Davies to T. J., dated 23 March 1921, obviously referring to the Shakespeare Head Press, asks,

'Can't we buy the beautiful type you said you could get for the Welsh University Press? If we could get a really good type we could have a nice little electrically driven press like the one in the National Library: the electric plant is on the spot and we could fix it up in the laundry. There are lots and lots of things we can do, and I am sure the beginning is the worst of it.'

Meanwhile, Gwen corresponded with Bullen's former secretary concerning the possible purchase of the Press.

By 27 April she was writing to T. J. to report that she had consulted John Owen, an uncle of the sisters who was also their financial adviser. He was not encouraging about setting up a press at Gregynog and advised that someone should report on the state of the Shakespeare Head Press and its equipment before buying it. She continued,

'Had a report of our respective accounts . . . no more pictures, etc. for ages! Both of us (especially I) terribly in debt!'

T. J. obviously offered to act as emissary in the negotiations for on 1 May she wrote again to him concerning the possibility of acquiring the goodwill of the Press but begged him not to make any firm promises, as she and Daisy had discovered they were liable for a payment of £29,000 each to a trust for improving housing in Wales, which they had overlooked. T. J. visited Stratford later in May but learned that, under the terms of Bullen's will, the Press was to remain in Stratford. To maintain a press so far away from Gregynog was clearly impracticable, but T. J. suggested that should the plant be sold the Davies's might make an offer for it. Nothing came of the venture for the Press was ultimately bought by Basil Blackwell and his partner who maintained it in Stratford until 1930 when it was removed to Oxford.

Gregynog was destined to become, *inter alia*, a conference centre. The first conference to be held was that of the National Council of Music in July 1921 under the chairmanship of Walford Davies[8] who since 1919 had held the Gregynog Chair of Music at the University College of Wales, Aberystwyth. Walford Davies became a close friend of the sisters and a staunch ally in many of their Gregynog activities. In October he stayed at Gregynog again to organise a weekend music retreat for staff and students from the College.

Plans were afoot to furnish the house, and also investigations were made concerning the possibility of including furniture making among the proposed crafts. Peter Waals, the well known furniture maker of Chalford in Gloucestershire, formerly employed by Ernest Gimson, was summoned to Gregynog in early June when he carried out a four-day feasibility study. His report was far from encouraging: while the estate sawmill was found to be in good working order, Waals considered the local craftsmanship rough and well below the standard of that in the Cotswold villages. As a result of the report the idea of furniture making was abandoned. Waals was also requested to submit designs for house furniture. These were based on Gimson designs and found favour with the sisters. After a visit to his Chalford workshop in October, an order was placed. Reporting the incident in a letter to T. J., dated 9 November 1921, Gwen Davies wrote,

'I love sun and colour and beautiful things and want others to share them too, and I can't help wanting to make Gregynog a home *and not an institution.*'

It had originally been intended to include pottery and weaving among the crafts envisaged, but by 1921 the sisters had become disheartened by the slow progress made by their architect in producing plans for the conversions necessary to the coach house and stable block at the rear of the house. This situation, together with Maynard's obvious lack of enthusiasm for some of the crafts, led to an eventual narrowing of their aspirations. In November Sir Edwin Hall relinquished his appointment, and, from then on, alterations were in the hands of T. W. Hughes, who was to be the sisters' estate manager for many years. By June 1922 the main structural alterations were completed. These included premises for a print shop and composing room. Now the idea of printing was firmly established, never to be dislodged.

Maynard took up his position at Gregynog on 6 July 1922 and by August was recommending the provision of an artist's studio which was subsequently built above the composing room.

Gwen Davies was in favour of appointing a librarian–secretary for the house and suggested to T. J. the name of Dora Herbert Jones, the friend who had assisted her in France and who was now a young widow, living in Aberystwyth and working as part-time palaeographer for the National Library of Wales. The idea, however, met with little enthusiasm from Daisy Davies, so the project was shelved, to be taken up again later in the history of the Press.

During 1922 Walford Davies had arranged three more musical gatherings at Gregynog. These were a working party from University College Aberystwyth, which met in April to prepare a school and college hymnal; the third annual conference of the National Council of Music, held in late June and early July; and a three-day conference on music in schools which was held in November.

Opposite: Insert from Christmas Card, 1922, with engraving by Robert Maynard.

A ddarlleno, ystyried;
A ystyrio, cofied;
A gofio, gwnaed;
A wnêl, parhäed.

GREGYNOG. NADOLIG. 1922.

On 20 December Maynard recorded in his notebook the pulling of the first proofs, printed on the Albion hand press which had been made specially for Gregynog by Stephenson Blake Ltd. On 27 December, 120 copies of the first Christmas greeting card were printed. The card is a modest but laudable first attempt on the part of one who had undergone no formal training in typography or printing and who was still teaching himself the art of wood engraving. It is printed in 14-point Kennerley type and consists of a single sheet of Japanese vellum inserted in blue-grey paper covers. On the front cover appear the lines by Ellis Wynne,

'Let him who reads, ponder;
And pondering, remember;
And remembering, act;
And acting, persevere.'

The imprint below reads,

'Printed at / THE GREGYNOG PRESS. / CHRISTMAS. / 1922.'

The insert bears a heavy and somewhat mechanical wood engraving depicting a view of the hall from a nearby hillside, signed by Maynard, underneath which is a Welsh translation of the quotation on the front cover and the words, 'GREGYNOG. NADOLIG. 1922.' That Maynard still had much to learn is evident from the inclusion of full stops printed at the ends of lines in capital letters. The sisters' reaction to this, the first piece of Gregynog printing, is not recorded. Cards bearing Christmas and New Year greetings from Gwen and Daisy Davies and their step-mother, and later from the Press to its customers, friends and well-wishers were thereafter printed every year until 1938.

The final event of the year was a Christmas house party. The fifteen guests included T. J. and his wife, René, Walford Davies, Sidney Greenslade and Professor and Mrs. H. J. Fleure.[9] Amidst music making and other festivities lengthy discussions took place on the future of Gregynog. Suggested activities included exhibitions of paintings, lantern lectures for school children, a summer school for Welsh art teachers who were to live a simple life, during the course, in accommodation provided over the stables, and the production of publications on art for school children, to be printed at the Gregynog Press. It was reported that Maynard was anxious to start the Press early in the new year.

Shortly afterwards Gwen Davies had further thoughts on the Press which she aired to T. J. in a letter. They must begin in a small way, she averred, limiting themselves to small editions which should be not only specimens of fine workmanship but also of literary value. Then, if Maynard were to prove successful, they could launch out into something more ambitious. She went on to remind T. J. that, some months previously, he had promulgated the idea of commissioning a special type face to be designed and cut for exclusive use by the Press and accused him of having lost interest in this in particular and of the artistic aspect in general. Poor T. J.! It is amazing that his interest remained so enthusiastic and his support so unwavering when one considers the multifarious activities in which he was involved in his official capacity as Assistant Secretary to the Cabinet. She continued,

'What worries me is that we are weak on the literary side, and I fear this is going to hamper us in the printing. Of Welsh stuff I have no doubt we can get plenty, but we must print English as well if we are going to employ such men as Maynard and Mason.'[10]

This is the first reference to John Mason, son of J. H. Mason, who was to become Maynard's assistant and bookbinder.

NOTES

1. 'Mr. Tom Jones . . . was . . . not only one of the most able and experienced, but also one of the most powerful servants of the state; a man of the utmost integrity and of a mind naturally generous and well-disposed towards his fellows.' Sir Osbert Sitwell's *Laughter in the Next Room*. Macmillan. 1949.
2. The National Library of Wales, founded in 1907, was at this time housed temporarily in the Old Assembly Rooms, Aberystwyth, now part of the university students' union.
3. Sir Edwin Thomas Hall, F.R.I.B.A., 1851–1923.
4, 5 & 6. Reminiscences by R. A. Maynard sent to T. J. on 3 March 1953.
7. J. H. Mason, 1875–1951. Assistant to T. J. Cobden-Sanderson at the Doves Press, 1900–1909. Assisted Count Harry Kessler to establish the Cranach Presse at Weimar in 1913 and resumed work there 1928–1929. Co-founder with Gerard Meynell and Edward Johnston of *The Imprint*, 1913.
8. Sir Henry Walford Davies, C.V.O., 1869–1941. Composer. Knighted in 1922. Appointed Master of the King's Musick in 1934.
9. Fleure was at that time Professor of Geography and Anthropology at University College Aberystwyth.
10. Undated letter from Gwen Davies to T. J.

4
First Productions

IF THE TWO PRECEDING YEARS ARE REGARDED AS the period of gestation, 1923 might be considered the year of experiment and partial fulfilment of some of the ideas promulgated by the Davies's and their close friends and associates.

Activity in the Press was gathering momentum. Maynard had no ambition to follow the tradition of hand printing employed by earlier private presses, and on 21 February a Victoria platen press was installed. This, a copy by Huddon of the better known German machine, was to do yeoman service, for on it was printed almost everything which subsequently came from the Press. Early in the year, two significant appointments were made. On 15 January, John Hugh Jones, an Anglesey boy straight out of grammar school, joined the staff as general assistant on a month's trial and was thereafter retained on a year's apprenticeship. On 20 March, John Mason was appointed bookbinder and general assistant. Mason, previously manager of Eyre & Spottiswoode's Eton College branch bindery, had agreed to come to Gregynog at a salary of £156 per annum plus free board and lodging in a flat above the print shop.

May 14 saw the completion of the printing of 2,000 copies of an account of the origin of the annual *Good-Will Message of the Children of Wales*, together with the text of the first broadcast message itself, which was issued on behalf of the Welsh Council of the League of Nations Union. Here Maynard, whose name appears for the first time in the colophon, experimented with the use of two colours, printing the text in black and the initials and paragraph marks in red. He was still working hard in an attempt to perfect his wood engraving technique, and in June appeared a single leaf, printed on hand-made paper, bearing a wood engraved mountain and lake scene, in black and pale blue, underneath which is J. Ceiriog Hughes's two stanza poem, *Rhyddid* (Freedom) in Welsh.

By the beginning of July the artist's studio was ready. This provided for Maynard the necessary peace and privacy to carry out his design

Opposite: Title-page to Poems by George Herbert, *printed in red and black.*

POEMS

BY

GEORGE HERBERT

THE GREGYNOG PRESS
MCMXXIII

work. The printing of the first book, *The Poems of George Herbert*, began on 3 July. Later the same month, the Press were also engaged in printing two four-page programmes for concerts given on 14 and 16 July, during the conference of the National Council of Music. These bear the earliest Gregynog Press device, a wood engraved GG[1] monogram imposed on the outline of a fir-clad mound.[2] The Press later used twenty-nine different devices.

The first fruits of the Press, *The Annual Good-will Message of the Children of Wales* and the 1922 Christmas card, were exhibited at the National Eisteddfod, held in Mold, Flintshire, 6–10 August, where they excited a considerable amount of interest. In addition to printing and designing, Maynard was engaged in another activity. The whole of August was devoted to running a summer school for art teachers at Gregynog, during the course of which an exhibition of paintings and sculpture was mounted, for which the Press printed a four-page catalogue. It had been intended that similar summer schools should be held annually, but Maynard was bitterly disappointed with the attendance (only seven students, drawn from Aberystwyth and Swansea), and after 1923 the experiment ended. Though the date is unrecorded, it seems likely that it was during the summer of 1923 that Maynard experimented with making pots, using clay dug from the estate. These were successfully fired at the South Kensington School of Art which, at least, proved the local clay suitable. One or two of the pots are still in existence at Gregynog, but for reasons unknown, possibly due to Maynard's lack of interest, no more pots were made. Indeed, from this time on there was no further mention of craft activity, apart from printing, at Gregynog. The vision, as originally conceived, had receded.

The printing of the three hundred copies of *The Poems of George Herbert*, selected by Sir Walford Davies, was completed on 9 November, though it was not published until 7 December. It was printed on the Albion hand-press in 12-point Kennerley type, on undamped hand-made paper made by Grosvenor Chater, and has a wood engraved frontispiece of Montgomery Castle, Herbert's birthplace, and initials by Maynard. Two hundred and fifty-seven copies were bound by Mason with a cloth spine and paste-marbled papers covers (for which the marbling was carried out at Gregynog), the remaining forty-three copies being bound at a later date in variants of either red or blue levant morocco. The colophon records that it was printed by R. A. Maynard, with the assistance of John Mason and John Hugh Jones. This chaste little book, unremarkable in itself, nevertheless, was a praiseworthy performance on Maynard's part. The weakest features are undoubtedly the initials which are somewhat uneven in size and are printed in a red ink which does not quite match the red used for the title-page device. The device itself is the same as that used earlier for the two concert programmes and is a little too large for the page.

Almost a week after the completion of the Herbert, Maynard printed one hundred and sixty copies of a four-page pamphlet addressed to the Council and Senate of Birmingham University, by the Geographical Association, which pointed out that Birmingham was then one of the only two universities in Britain where geography was not recognised as a subject worthy of study and research. The pamphlet was printed at the request of Professor H. J. Fleure of Aberystwyth, currently Honorary Secretary of the Geographical Association, a close friend of the Davies sisters who was frequently called in to discuss Press matters in the early years.

The final piece to come from the Press that year was a small Christmas card from Mrs. Edward Davies and her two step-daughters. Inside a blue-grey cover is a double fold bearing two verses on the verso, facing the greetings message which is surmounted by a small Maynard wood engraving depicting the front of the family home, Plâs Dinam. A small insert invites recipients to Christmas music and carols on the evening of 29 December, in the Music Room.[3]

Eighteen guests were present at a post-Christmas house party, held 28–31 December, and these included Dora Herbert Jones, already well known as a collector and singer of Welsh folk songs, who entertained the assembled company. A 'Daddy' Rothwell organ had recently been built and installed in the Music Room, and a new Broadwood grand piano had been delivered in time for the festivities. Not all was mere jollity, however, for there were further discussions relating to the future of the Press.

NOTES

1. GG = Gwasg Gregynog = Gregynog Press.
2. The inspiration for this sprang from Celynog, a fir-topped hill in the vicinity of Gregynog.
3. The Music Room had fairly recently been created from the former billiards room of the house.

5

Early Books

THE PRESS STAFF WAS FURTHER AUGMENTED IN February 1924. On the twentieth, Horace Walter Bray, an old friend and erstwhile fellow student of Maynard, then working as a London scenic artist, was appointed as resident artist. Bray had no knowledge of wood engraving but was swift to learn the technique from Maynard and, for the next seven years, was to share with the controller the design and execution of virtually all the illustrations and initial letters used in the Press. Another appointment made that month was that of Idris Jones, a local boy, as general assistant. Jones later became the Monotype caster operator and printing assistant. Finally, he succeeded Herbert Hodgson as printer, a post which he held until the outbreak of war in 1939.

Negotiations had been under way for some time to produce an edition of Ceiriog's poems, edited by J. Lloyd Jones,[1] and printing began on 21 February. Maynard was gradually gaining in skill and confidence as typographer and printer. He had obviously settled happily into his new vocation, for on 12 March he wrote to T. J.,

'The press means so much to me now that I am, naturally, impatient at anything which comes to interrupt what I think of as my real work here.'

Some years before his death, in sending his reminiscences to T. J., Maynard was to recall those early years in the Press as the happiest of his working life.

In March, T. Gwynn Jones, poet and Professor of Welsh Literature at University College Aberystwyth, offered the Press Board a collection of his poems for publication, an offer eagerly accepted. (The Press Board, then in its infancy, consisted of the Davies sisters and T. J. only, but advice was frequently sought from close friends and associates. Maynard acted *ex officio*.) In answer to a request that the book should be dedicated to T.J., the poet received a flattering reply,

'I can truthfully say that to appear on your page will rejoice me more than any honour His Majesty's government could give me.'[2]

In 1924 a great change was effected in the lives of Gwen and Daisy Davies. Their brother, David, after his second marriage in 1922, had lived first at Broneirion, his grandfather's home, which lay across the river Severn from Plâs Dinam but, at the wish of his wife, had moved into Plâs Dinam in 1923. The move led to a difference of opinion between David Davies and his sisters who, early in the spring of 1924, revealed to T. J. their intention of leaving the family home and removing themselves to Gregynog. This was the first time that they had contemplated Gregynog Hall as a potential home. Now it was decided to reserve part of the Hall for their own private use; building alterations were swiftly put in hand, and the removal to Gregynog was effected in mid-April. Their art collection went with them.

Poems by George Herbert, published the previous winter, was well received in discerning circles. Alfred W. Pollard of the British Museum was among the first to congratulate the Press in writing. On 16 April, T. J. dined with C. H. St. John Hornby, Sir Charles T. Hagberg Wright and Henry Yates Thomson, three well known bibliophiles, all of whom spoke warmly in the book's favour. The problem was what to print next, after the books already in preparation. Gwen Davies suggested that T. J. should seek the advice of Ernest Rhys, editor of Dent's *Everyman Series*, with whom he had recently become acquainted. Rhys, only too willing to help, was summoned to spend the weekend of 7–10 June at Gregynog to discuss plans for future publications. A memento of the visit, which was to be the first of many, was recorded by Rhys in the visitors' book,

'When ideas failed us, we went and stared at the Turners, or the Botticelli's . . . or at Whistler – or at the riot of rhododendrons and azaleas in the June rain; – or the ladies of Gregynog gave us music in the Music Room.'

Later the same month there was an even more distinguished visitor to the Hall. Sir Edward Elgar, having attended a performance of his work, *The Apostles*, at Harlech motored back to Gregynog with the sisters and there spent the night of 27 June.

Meanwhile, in June, Maynard was busy printing a poem entitled *Written on Walking in the Woods of Gregynog in Montgomeryshire* which was first published in 1794 in *Poems on Several Occasions*, by Mrs. Darwall, formerly Miss Whateley. Above the eight stanzas appears a wood engraving by Maynard, depicting the lawns and woods of Gregynog.

Gwen Davies wrote frequently and copiously to T. J. concerning

Press matters. A letter, dated 4 July, demonstrates that the Board was still trying to formulate objectives. Concerning future policy, the letter includes the following notes for T. J.'s consideration and comment:

'1. The primary objective in our minds at the present is to unlock the treasure house of Welsh literature, romance and legend, and make it accessible to the English-speaking public.
2. To present it in as perfect a form as we possibly can as regards the English translation and artistic treatment [so] that it must inevitably attract attention and win for the little country the tardy recognition of cultural nations which I feel certain has only been withheld because so few can understand the Welsh language.
3. Further, to collect and reproduce the *best* work that Welshmen have contributed to the *vast* store of English literature.'

The letter continues,

'We must bring to this task a strong critical sense and permit nothing which is not of *the best;* also much patience and foresight are needed. As regards the Welsh books, I feel that you are afraid there will be little or no demand for them. This does not worry us in the very least. It is worthwhile producing a beautiful thing for its own sake. We consider certain Welsh things are worthy of a beautiful setting . . .'

T. J. had recommended that some of the Press books should be given away as school prizes, but the idea was given short shrift, as this letter later shows. Gwen feared that a fair proportion so donated would go to unappreciative recipients. She also pointed out that, as school prizes, they would need to be bound in full leather, as was then the custom. This would add four-and-sixpence to the cost of the book for a full niger binding or seven-and-sixpence for one in full levant, plus the cost of lettering in gilt. Cost was only one consideration. Furthermore, such bindings would monopolise Mason's time. To send out the books for binding would be tantamount to defeating the objects of the Board. On the other hand, to increase the bindery staff would necessitate the provision of additional housing. She and Daisy were agreeable, however, to donating a copy of every Press publication to each college in Wales and to each Welsh member of the Council of the National Library of Wales. They were also prepared to sell the Welsh books at one half their cost price. Gwen pointed out that, in future, the Welsh books need not have as many illustrations as the Ceiriog poems currently in preparation.[3]

Two constantly recurring themes in Gwen Davies's letters show her to have been tormented by the fear of the possible depletion of their capital and frustrated by her desire to give something to her fellow men other than mere money. The letter, already referred to, gives ample evidence of these anxieties:

'We do not want to make any personal profit out of the books, but would like to feel that some day the Press will be able to pay its own way[4] . . . so that if anything should happen to us, the Press would still continue.'

Later, in the same letter, referring to the proverb that, while one may lead a horse to the water, one cannot make it drink, she continued,

'He the Welshman is so self-complacent, so self-sufficient, so ignorant – how are we to convince him that he is thirsty? Oh, T. J. we have so much to give that isn't just money, money and yet more money . . .'

The National Council of Music met for its annual conference at Gregynog in July, and once again Walford Davies organised and conducted a concert given by staff and students from Aberystwyth, accompanied by a contingent from the Welsh Symphony Orchestra. For this concert Maynard printed a two-page programme. Otherwise, work in the Press was concerned largely with the preparation of *Caneuon Ceiriog* and an edition of Henry Vaughan's poems which Ernest Rhys had agreed to select and edit.

In July Ernest Rhys was consulted by letter concerning the pricing of books. His reply, however, was not a practical one, concerned with pounds, shillings and pence:

'I think it is most important at first to create beautiful books, representing the world of Welsh literature and make the great outer world aware of them. Out of this will grow the true Gregynog tree.'[5]

Maynard, following private press tradition, was keen to acquire a typeface peculiar to Gregynog. He had recently studied Stanley Morison's *Four Centuries of Fine Printing*[6] which reproduced more than six hundred specimen pages from books printed 1500–1914 in roman founts, but opined that none of the types was entirely suited to Gregynog's needs:

'The nature of the books that we are planning to print demands a type holding at least some hint of medievalism: some trace of Celtic inspiration.'[7]

When the matter was discussed by the Press Board in September, it was decided that Edward Johnston[8] was the obvious person to approach. Maynard was instructed to write to Johnston enquiring whether he would be prepared to design a new type on behalf of the Press. Maynard wrote,

'The nature of these books, their Celtic source, and the fact that we are desirous of introducing versals and woodcut decorations, incline us to the idea of having a type for our own exclusive use.'[9]

Johnston's reply, which came by return, brief and to the point, was less than encouraging,

'I am rather loathe [*sic*] to design a new type until I see better use made of what we have – or a real need for it.'

At the time at which it was written, when the Monotype Corporation's programme for reviving classic older typefaces, which was to have such far reaching and beneficial influence on British printing, was still in its infancy, Johnston's reply made sound sense, even if it appeared a little harsh in its judgment. Gwen Davies, however, clearly remained unconvinced. Whilst *The Life of Saint David* was still in the planning stage, she raised the question of a typeface once again, in a letter written to T. J. in October. This latest book, she considered, ought to be printed in a special type, the design of which should be influenced by the old Welsh manuscript hands. She was obviously overruled, for the book was eventually set in Poliphilus. It was not until 1934 that a special type was made for the Press.

The autumn passed quietly, the Press staff now at work on two books and Gwen Davies penning long missives to T. J. concerning Press matters. The form of address, which had hitherto been 'Dear Mr. Jones', changed in a letter written on 31 October which began,

'Dear T-J,
I really cannot behave properly any longer, I'm sorry!'

A letter of 3 November comments, in reference to some pending decision,

'I do not like doing things without Daisy, for she has just as much share in the Press as I have.'

For all that, the share was mainly financial. Though Daisy Davies did sit on the Press Board and was consulted frequently on Press business, there is little evidence from the correspondence, or in the memory of those who knew her, of any burning intrinsic interest on her part. Increasingly the Press became the brainchild of T. J. and Gwen Davies. Ultimately, all major decisions were theirs, arrived at by common consent, born of a shared enthusiasm and mutual respect.

Poems by Henry Vaughan, selected by Ernest Rhys, which had been completed in the autumn, was published early in December. Despite Gwen Davies's private complaint to T. J. that Rhys's thirty-page introduction was too long and too personal, the piece stood. The book is very much a companion volume to *Poems by George Herbert*, to which it is similar in all respects. The woodcut initials are again somewhat overlarge, with the result that some of the wider initials, appearing on the recto, run too far into the gutter. The binding, with dark blue cloth spine and blue-grey paper-covered boards, is printed on both covers with the Swan of Usk, the poet's dates and a monogram of his initials surmounted by an S (Silurist). The first thirty copies were later bound in variants of crimson and blue levant, plain except for a single gilt-rule panel, containing on the front a gilt-blocked swan.

Work in the Press closed, as usual, that year with the printing of a Christmas card, bearing greetings from Mrs. Edward Davies and her two step-daughters. Blue-grey covers carry a wood engraving of a tree, incorporating the word CHRISTMAS and the date. The greetings message appears on the first recto of a four-page insert, of which the two middle pages are devoted to five verses of a Christmas carol. At the head of this is a small circular wood engraving, depicting a shepherd and the Star of Bethlehem.

Nineteen-twenty-four ended with a house party and a meeting of the Press Board, held on 29 and 30 December, at which it was agreed to defer any decision concerning advertisement of the books and to offer the Herbert, the Vaughan and the forthcoming Ceiriog volumes together to subscribers and to the trade. Maynard took the opportunity to lay several requests before the Board. The first of these was that a Monotype casting machine be purchased to enable a wider range of types to be used. The Board gave permission for the order to be placed immediately.[10] The other main request was for additional staff—a compositor, a pressman, a bindery assistant and a clerical assistant. Approval was granted in a limited way. As the appointment of a compositor was deemed to be the most urgent, the Board agreed to appointing someone in three months' time. Maynard was requested to seek out and train a local girl for the bindery as it had earlier been decided to recruit local staff wherever possible. It was considered that clerical assistance should be obtained on a part-time basis. No decision was made concerning the appointment of a pressman. Once again the point was made that additional staff would involve not only more money for salaries but also the provision of additional housing. Maynard and Bray and their respective families were already accommodated, Maynard in Bronbechan, a former coachman's house, in the neighbouring village of Tregynon and Bray in Tŷ Canol, a house recently built for his use in the same village.

Over the years the Davies sisters amassed a collection of early and fine printed books for their library at Gregynog. Ernest Rhys appears to have been making purchases in London on their behalf, for in January 1925 Gwen begged T. J. to restrain Rhys, as they wished to choose their own books!

Gregynog was remote from any book-producing centre and, in order to provide some stimulus, Gwen Davies decided to foot the bill for some exhibition visiting. In May, Maynard visited a book fair in Paris and the Second International Book Fair held in Florence. T. J., accompanied by Ernest Rhys, was also despatched to the latter but, with the exception of a few German and Italian exhibits, the travellers declared themselves disappointed in what they had seen.

During their absence, the first book in Welsh, *Caneuon Ceiriog: Detholiad*, edited by Professor J. Lloyd Jones, was published. For this, 14-point Kennerley type was again used. The spacing is good, and the type sits well on the page. The woodcut initials show an improvement on those in the two earlier books and accord well with the thirty-one simple, chunky wood engravings, designed and executed by Maynard and Bray, most of which form head- and tail-pieces to the text. The book was bound with a white linen back, with printed title-label and grey paste-marbled boards. The binding of the first thirty copies, designed by Maynard, again in variants of red and blue levant, has a gilt pattern of light bordering lines, with small corner leaf sprays. The book was sympathetically reviewed in *The Times Literary Supplement*[11] and *The Observer*,[12] but in Wales the sales were discouraging, and it was some years before the edition was sold out.

John Mason had grown disenchanted with Gregynog, partly due to the failure to provide him with the house which had been promised, and, late in the summer, he tendered his resignation. In a bitter letter to T. J., dated 6 July, Mason aired his grievances,

'I doubt very much whether the conditions which underlie the external appearances of Gregynog would allow any skilled craftsman that state of mind that is necessary to creative work. He would, at any rate, need to have a very different temperament to mine.'

This was the first sign of staff discontent and an indication of the state of friction which came to exist between T. W. Hughes, the estate manager, and the majority of the Davies employees. Mason left the Press later that month to take up a post at the Shakespeare Head Press.

Ever since January, T. J. had been attempting to persuade Dora Herbert Jones to travel over from Aberystwyth on one day each week to assist with secretarial duties. She finally agreed to come in September, provided she could find someone to look after her two young children. She eventually came and remained as Secretary in a part-time capacity until she moved to Tregynon in 1927.

During the summer the works of the Press had been exhibited at the National Eisteddfod, held in Pwllheli, where they were admired by J. Gwenogvryn Evans[13] who had been responsible for publishing privately some of his own translations of early Welsh manuscript works. In September he wrote to the Davies's offering material for publication by the Press,

'A portion of the Gospels . . . translated into Welsh about 1300. I copied the manuscript some years ago and was wondering whether you would feel inclined to print it . . . There is also an old translation of the Paternoster and a few verses besides. I have lost over £300 on Aneirin,[14] and in consequence I don't wish to burn *my own fingers* any more just now! Your fingers are better gloved than mine . . .'[15]

Nothing came of his offer, but the incident gives some indication of the growing reputation of the Press.

On Saturday, 31 October, a concert was given at Gregynog by the Aberystwyth Small Choir, under the direction of Walford Davies. The Press printed an eight-page programme which is of interest as it displays the first use of a type other than Kennerley, made possible by the purchase of a Monotype caster earlier in the year. The type used was Garamond, issued by the Monotype Corporation only in 1922 and which did not have its first commercial showing until *Penrose's Annual* for 1923.[16] The concert was attended by Adrian Boult who had first visited the Hall in May when he had conducted the orchestra for the Montgomeryshire Music Festival in Newtown, an event which had been started with the generous support of the Davies sisters. Thereafter, Boult was to be a frequent visitor and later conductor of the Gregynog Choir[17] for the festivals of music and poetry.

The Press had lacked a binder since Mason's departure in July.

Sydney Cockerell, the son of Douglas Cockerell, came as a temporary stand-in for two months in mid-September. While at Gregynog, he worked in the bindery and also carried out further experiments with marbling paper, a craft for which he later became renowned. After Cockerell's departure, paper marbling was discontinued.

In an effort to replace Mason, T. J. had consulted Douglas Cockerell who had recommended contacting George Fisher, one of his past pupils. Fisher, a Londoner, had learnt forwarding with an amateur bookbinder and had subsequently been apprenticed as a finisher at Rivière's, the fashionable West End bindery. In 1911 he had bought a five-acre smallholding at Medstead in Hampshire and, from that time on, had devoted most of his time to farming. In 1925 he had done no book binding for ten years. T. J. wrote to Fisher inviting him to come to Gregynog on a month's trial. Farming was at a low ebb in the mid-1920's, and times were hard for Fisher, so that he finally agreed, albeit reluctantly, to the proposal. He arrived at Gregynog about 18 November. Maynard and the Press Board were greatly impressed with the standard of his work, and when he returned to Medstead, after the trial month had expired, he was invited to join the staff on a permanent basis.

Fisher was in a dilemma: debts had accrued gradually over the last few years, and small farms were virtually impossible to sell. He wrote to the Press Board stating that he would be unable to return unless the salary offered was drastically increased, and quoted a figure far in excess of anything likely to meet with the Board's approval, but, by return, came the offer of a contract! So it was that Fisher came to Gregynog where he was to remain for the next twenty years, becoming the longest-serving member of the Press staff. His farm was never sold: his wife continued to run it, and Fisher returned each summer to help with the hay harvest.[18]

By November, the Press Board was much occupied with plans for producing the first Press prospectus. A draft was prepared and sent to Ernest Rhys for comment. Rhys wrote to say that, having discussed the draft with Douglas Cockerell, he was of the opinion that it lacked force. He suggested that it should be re-written and offered to write a descriptive piece about the Press for inclusion in the revision. A small, but characteristic, alteration was made to the first draft by Maynard who reported that he had deleted the word *special* which had been used to describe the Ceiriog illustrations:

'It is a word I strongly object to in connection with any artistic effort.'[19]

One thousand copies of the revised prospectus, set in Garamond type, were printed on 8 December. It comprised a one-page description of the Press by Rhys, together with a two-page announcement of the first three books and five specimen pages. It included also a new, large, handsome Press device, designed by Maynard.

The 1925 Christmas card was a simple four-page affair, printed in red and black and decorated with a wood engraving which is an enlarged and somewhat embellished edition of the earliest Press device. The greeting, for the first time, was worded impersonally,

'To bring greetings and best wishes for nineteen twenty-six from Gregynog'.

This was obviously designed as an all-purpose card which could be sent out by the Davies sisters or by the Press to subscribers and friends.

The permanent appointment of Dora Herbert Jones, who had been working one day per week since September, was sanctioned at a Press Board meeting held on 30 December, following the usual post-Christmas house party. For a salary of £150 per annum she was to undertake secretarial duties to the sisters, take charge of the library, undertake all clerical duties and accounts for the Press, check proofs in Welsh, as required, and act as Secretary to the Press Board. The appointment could not come into immediate effect as Maynard was still living in Bronbechan, the house intended for her use, pending the building of a new house a little further up the lane.

Early in 1926, Maynard began to show signs of resentment against T. W. Hughes, the estate manager, whom he accused of interfering in relations between him and members of the Press staff. This was the beginning of a long war of attrition between Maynard and Hughes which continued unabated until Maynard left the Press in 1930.

In April the promise of a compositor was fulfilled by the appointment of Richard Owen Jones. Jones, who had worked as compositor on a local Newtown newspaper, was a skilled craftsman. During the next eleven years he was to make a significant contribution to Press works in the immaculate composition which has always been a much admired feature. Dora Herbert Jones later recalled being fascinated by watching him patiently spacing letters, using pieces of thin paper, to achieve a result which would satisfy his critical eye.

The following month saw the completion of T. Gwynn Jones's *Detholiad o Ganiadau*.[20] Reviews were mixed. For example, Ernest Rhys, who reviewed it in August for *The Manchester Guardian*, found the paper too yellow, the type over-large for the page and the illustrations not entirely harmonious. One must admit that, though competently printed (by Horace Bray and Idris Jones), it is a dull looking piece, compared with other Gregynog books. The author, however, was well pleased with the production, as a letter to Maynard reveals,

'I very deeply appreciate the honour of having my verse thus issued at the hands of artists and gentlemen . . .'[21]

The ordinary binding is in full peacock-blue buckram—an unfortunate choice as this has faded badly with the passage of time. Bray was responsible for the design of the levant binding, executed in variants of red and green, which is decorated with a trellis work of blind lines, connected across the spine with a gilt hinge pattern.

The prospectus for this book is interesting in that it announces:

'The Press will undertake the binding of a limited number of copies in full levant morocco. The price of the volume thus bound will be two and a half guineas.'

(The price of the ordinary edition was fifteen shillings.) The bindings, hitherto cased by hand in cloth and paper,[22] obviously provided no outlet for Fisher's skill, and it was at his instigation that a limited number of copies of each edition came, from this time on, to be bound in full leather, forwarded, covered and finished by his own hand. These bindings, now prized collector's pieces, were to add lustre to the reputation of the Press.

On the death of the Davies's uncle, John Owen, Major W. J. Burdon Evans, a barrister, was appointed as their financial adviser. As seemed only natural, he also assumed responsibility for handling the finances of the Press and, in consequence, was made a member of the Press Board.

The League of Nations Union Welsh National Council's Advisory Education Committee held its first Gregynog conference 15–18 October. For this the Press printed a sixteen-page programme in Poliphilus type, for which they had just obtained the necessary matrices. The programme also includes a small wood engraving of Gregynog Hall. Another ephemeral piece followed swiftly on its heels. This was a single-fold programme for a concert of vocal and chamber music, given on 30 October under the direction of Walford Davies, for which the Press reverted to Garamond.

At the suggestion of Ernest Rhys, Edward Garnett had been invited to contribute a prospectus note and foreword to *Chosen Essays by Edward Thomas*, selected by Ernest Rhys, the printing of which had been completed in August. The copy for the note, which includes the name of Charles Doughty,[23] was written in Garnett's own hand, a hand not always easy to decipher. Maynard was obviously unfamiliar with

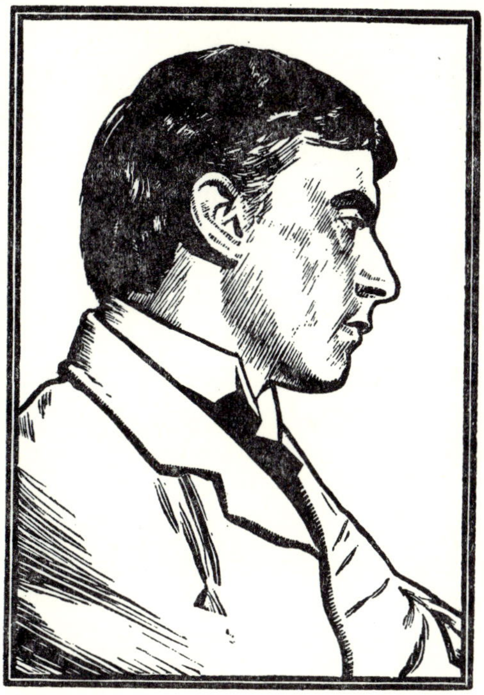

Portrait of Edward Thomas engraved by Robert Maynard
from Chosen Essays by Edward Thomas.

Doughty's name, with the result that it was mis-spelt as *Daughty* in the proof sent to T. J. for his approval. In reply to T. J.'s gentle reproof, Maynard blamed the inadequacy of the Gregynog library as a source of reference,

'People like Rhys and Garnett should not rely on us to correct their mistakes. The Gregynog library is more than disgraceful. They haven't a single book of any use.'[24]

The book was published on 10 November and was well received by the critics. *The Times Literary Supplement* gave it a favourable review[25] but, like most reviews of Gregynog publications, it avoided all mention

YR EFENGYL SANCTAIDD YN OL SANT LUC

Pennod 2, Adnodau 6 hyd 21

A BU TRA'R OEDDYNT HWY YNO, gyflawni ei dyddiau hi i escor. ✠ Ac hi a escorodd ar ei mab cyntaf-anedic, ac a'i rhwymodd ef mewn cadachau, ac a'i rhoddes yn y preseb, am nad oedd iddynt le yn y lletty. ✠ Ac yr oedd yn y wlad hono fugeiliaid yn gorwedd allan, ac yn gwilied eu praidd liw nos. ✠ Ac wele angel yr Arglwydd a safodd ger llaw iddynt, a gogoniant yr Arglwydd a ddiscleiriodd o'u hamgylch, ac ofni yn ddirfawr a wnaethant. ✠ Yna yr angel a ddywedodd wrthynt, Nac ofnwch, canys yr wyf yn mynegu i chwi lawenydd mawr, yr hwn a fydd i'r holl bobl: ✠ Geni i chwi heddyw geidwad, yn ninas Dafydd: yr hwn ydyw Crist yr Arglwydd. ✠ A hyn [fydd] arwydd i chwi, Chwi a gewch y dyn bach wedi ei rwymo mewn cadachau a'i roddi yn y preseb. ✠ Ac yn ddisymmwth yr oedd gyd â'r angel nifeiri mawr o luoedd nefol yn moliannu Duw, gan ddywedyd, ✠ Gogoniant i Dduw yn yr uchelder, a thangnefydd ar y ddaiar, ac i ddynion ewyllys da. ✠ Ac wedi darfod i'r angelion fyned oddi wrthynt i'r nefoedd, yna y bugeiliaid a ddywedasant wrth eu gilydd, Awn hyd Bethlehem, a gwelwn y peth hyn a ddangosodd yr Arglwydd i ni. ✠ A hwynt a ddaethant ar frys, ac a gawsant Mair, ac Ioseph: a'r dyn bach yn yn gorwedd yn y preseb. ✠ Pan welsant, hwynt hwy a gyhoeddasant yr hyn a ddywedasid wrthynt am y bachgen hwnnw. ✠ A phawb ar a'u clywsant, a ryfeddasant am y pethau a ddywedase y bugeiliaid iddynt. ✠ Eithr Mair a gadwodd y pethau hyn oll gan eu hystyried hwynt yn ei chalon. ✠ A'r bugeiliaid a ddychwelasant gan ogoneddu a moliannu Duw, am bob peth a'r a glywsent, ac a welsent, fel y dywedasid iddynt. ✠ A phan ddaeth yr wythfed dydd i enwaedu ar y bachgen, galwyd ei enw ef IESU.

of the book's physical presentation. C. H. St. John Hornby wrote an admiring letter, and Edward Garnett praised it but complained that the title-page portrait, engraved by Maynard, did not resemble Thomas as he was in life. This, the fifth book, is altogether lighter in conception and is the first book to appear in a type other than Kennerley, being printed in 14-point Garamond. Most of the edition is on Dutch Van Gelder paper, but the first one hundred copies are printed on a Batchelor special make of hand-made paper, watermarked with a large GG monogram. The twenty-four wood engravings by Maynard and Bray, which are mainly head- and tail-pieces, show considerable improvement on their former work, having a much more lively quality. The ordinary binding, in blue buckram, has a gilt-blocked outline vase of flowers on the upper cover. The special binding, in light blue levant morocco, was designed by Maynard. It bears a trellis work of blind lines, studded with small gilt squares, which encloses two sets of concentric gilt panels, set one above the other. Fisher later recalled,

'This binding gave me much pleasure. Its plain, closely fitting squares demand much of the craftsman. The effect, to me, is one of coolness and clearness, like an early morning.'[26]

The 1926 Christmas card was the first piece of *fine* non-book printing to emanate from the Press. It consists of verses 6–21 from the second chapter of the Gospel according to Saint Luke, reprinted from the *Bishop Morgan Bible* of 1588, printed in Welsh on a single leaf, enclosed within grey-blue paper covers. The text is handsomely set in Poliphilus and printed in red and black on Batchelor hand-made paper. The item was much admired, particularly in Welsh circles, and rightly so.

NOTES

1. John Lloyd Jones, 1885–1925, Professor of Welsh, University College Dublin.
2. Letter from T. J. to T. Gwynn Jones, dated 5 April 1924.
3. The large number of illustrations (31) in the Ceiriog were attributed to the fact that Maynard and Bray were short of work and had time on their hands.
4. Which it never did!
5. Letter from Ernest Rhys to T. J., dated 26 July 1924.
6. Stanley Morison – *Four Centuries of Fine Printing*. Benn. 1924.
7. Letter from Maynard to T. J., dated 26 August 1924.
8. Edward Johnston, 1872–1944, illuminator and calligrapher. The suggestion to approach Johnston came from Stanley Morison whom T. J. had consulted.
9. Letter from Maynard to Edward Johnston, dated 8 September 1924.
10. There was no keyboard. The caster was used merely to produce type which was then hand-set. As Monotype metal was comparatively soft, the Press opted for

Opposite: Insert from Christmas Card, 1926, printed in black and red. (*Side margins reduced.*)

using a harder foundry metal with the result that frequent stoppages on the machine had to be endured.

11. Reviewed 29 April 1926.
12. Reviewed by Ernest Rhys, 27 June 1926.
13. John Gwenogvryn Evans, 1852–1930, a noted Welsh paleographer.
14. *The Text of the Book of Aneirin.* Privately published. Pwllheli. 1908.
15. Letter from J. Gwenogvryn Evans to the Misses Davies, dated 29 September 1925.
16. Printed in December 1922.
17. A small choir, comprised mainly of Gregynog staff, formed in 1928 principally to provide music for the services. It consisted of about twenty-eight voices, augmented to forty for festivals. Most of the choir were incapable of reading music, so that items were transcribed into tonic solfa by Gertrude Rowlands, Dora Herbert Jones's sister, and copied out by Dora and Hubert Davies, a member of the choir.
18. Fisher's wife did initially come with him to Gregynog, but after six weeks her health was already beginning to suffer from the damp climate, and she returned to Medstead.
19. Letter from Maynard to T. J., dated 13 November 1925.
20. There is obviously an error in Thomas Jones's bibliography, where the publication date is given as 8 January 1926, whereas the colophon stated that it was completed on 7 May.
21. Letter from T. Gwynn Jones to Maynard, dated 29 March 1926.
22. The levant binding of *The Poems of George Herbert* is by John Mason.
23. Charles Montague Doughty, 1843–1926, explorer and writer. Author of *Arabia Deserta*, 1885.
24. Letter from Maynard to T. J., dated 18 November 1926.
25. Reviewed in *The Times Literary Supplement*, 30 December 1926.
26. Fisher in conversation with the author, May 1969.

6

Reputation Established

IN FOUR YEARS THE PRESS HAD PRODUCED FIVE books. It was now firmly established as a viable unit with a growing staff. Its publications so far had been received with approbation in circles which mattered, and the scene was set for the fulfilment of its early promise. It was recorded, with some satisfaction, at a Press Board meeting held on 12 January 1927 that all 500 copies of T. Gwynn Jones's *Detholiad o Ganiadau*, published in January of the previous year, had been sold. Idris Jones had acted as pressman for the two books produced in 1926, but Maynard was still urging the appointment of a skilled printer. T. J. offered to consult the managers of the Baynard and Curwen presses with a view to finding someone suitable.

In March the Director of the National Museum of Wales enquired of T. J. whether the Press would be prepared to print a formal address to Their Majesties, King George V and Queen Mary, for the occasion of their opening the new museum building on 21 April. T. J. wrote, almost apologetically, to Maynard, saying that he assumed the Press would be too busy to undertake the task at such short notice. Maynard, however, realised the prestige to be gained from such an assignment and readily agreed to do it. The ten-page address is a handsome piece of bilingual printing, set in Poliphilus and Blado italic type and printed in red, blue and black, with a large wood-engraved opening and decorative initials. The binding is in blue levant morocco, decorated in gilt with a semée leaf and square dot pattern which surrounds a title panel on the front cover.

The renowned collection of bookbindings formed by the late Major J. R. Abbey was then in its infancy.[1] Captain Abbey, as he then was, wrote to the Press on 19 March to say that he had seen copies of the first two publications bound in full levant which he greatly admired. He had purchased copies of the ordinary edition and enquired whether the Press would consider re-binding those in levant also. This Fisher did, at a price of thirty-one shillings and sixpence for the *Poems by George Herbert* and two guineas for the *Poems by Henry Vaughan.*

The long-awaited arrival of Dora Herbert Jones as full-time Secretary to the Press occurred about 20 April. For the next ten years she was to prove a tower of strength, both to the Press and, privately, to the sisters. Yet her position was never an easy one. She had long been a friend of Gwen Davies, and they had many interests and tastes in common. Gwen came to rely on her heavily, from this time on, as personal assistant and confidante and sought her help with all the Gregynog activities. She took part in arranging events, in particular the religious services held during conferences and the festivals of music and poetry held later on. To be a friend, a ready help in times of need and yet, withal, an employee whose main task was to act as secretary, public relations officer and saleswoman to the Press, was to prove no easy task for a highly cultivated woman, with a mind of her own, who cared passionately about the maintenance of high standards. During the years 1927–1937 she came to put loyalty to the Press and its owners before anything else, save for the care of her children. Her position was particularly unenviable in later years, after Maynard's departure, when there were conflicts of personality within the Press. The ambivalence of her position meant that she was suspected by her colleagues of spying for the owners and that she was suspected by the Davies's of inspiring insurrection in the staff ranks.

Shortly after Dora Herbert Jones joined the permanent staff, the Gregynog Choir was formed. This was a small mixed choir composed mainly of employees, augmented by a few voices from the village of Tregynon. Its conductors were successively Sir Walford Davies and Sir Adrian Boult. Many years later, in a radio broadcast, Sir Adrian was to refer to the choir's rendering of Walford Davies's setting of the *One Hundred and Twenty-First Psalm* as the most perfect performance of that work he had heard.

Dora Herbert Jones's appointment was followed, three weeks later, by that of Herbert John Hodgson as pressman. Hodgson, a Londoner, had already had a varied and interesting career in printing. Not long before, he had printed the first limited edition of T. E. Lawrence's *The Seven Pillars of Wisdom*, a project which was financed and supervised personally by Lawrence.[2] Hodgson was recommended to the Press Board by C. H. St. John Hornby who had known him during his years in the printing department of W. H. Smith & Sons. An interview was arranged with Maynard in London at which Maynard cautiously engaged Hodgson to come to Gregynog on a month's trial, at an annual salary of £234 per annum. This initial agreement was transmuted in November into a three-year contract which was subsequently renewed at intervals until Hodgson left the Press in 1936. Despite earlier reservations, Maynard eventually came to be of the opinion that Hodgson was probably the best pressman in the country at that time.

Hodgson well remembered his introduction to Gregynog. On arrival at Newtown station, he hired a horse-drawn cab to take him to the Hall which he mistakenly believed to be in the town. Due to his cockney accent and inevitable mispronunciation of Gregynog, the driver was at a loss to know where he wished to be taken, but hopefully drove him round the town for fully half an hour before depositing him back at the station! The dilemma was solved when Ted Walters, the Davies's chauffeur, arrived on the scene to collect him.

The arrival of Herbert Hodgson and his family in the tiny village of Bettws Cedewain, some two and a half miles from Gregynog, where he had been assigned a cottage, must have caused a considerable stir, for his cockney ways were far removed from those of his neighbours. He was, however, to make a considerable contribution to the social life in the neighbourhood as he formed a small dance band which performed at functions in Bettws, Tregynon and in villages much farther afield. Hodgson himself played the mandolin-banjo and travelled to his engagements on the motor-cycle combination belonging to the group's drummer. As these village dances were normally held between 8 p.m. and 2 a.m., Hodgson and his drummer had many an exciting and perilous drive over the hills on winter nights when road conditions were bad. As they approached the village in the early hours of the morning, the engine was switched off and they coasted silently down to Bettws, so as not to awake their sleeping neighbours.

It was due to Hodgson's influence that George Fisher became a dance enthusiast also and followed the band in his small car, accompanied by his wife's niece, Jessie Adams, who was not only his dance partner but also his housekeeper throughout his long years at Gregynog.

While in London that May, Maynard, in company with Hugh Blaker, met Augustus John who had earlier expressed an interest in illustrating a book for the Press. John now suggested that he might do some wood engravings for a book of Welsh gypsy folklore, collected by John Sampson.[3] Though long familiar with etching, the artist knew nothing about wood engraving but saw no reason why he should not acquire competence in the medium, especially as Maynard offered to help him. Though Maynard kept his promise and sent John blocks, a set of gravers and a textbook, together with an offer of personal tuition, no illustrations were ever forthcoming. John was of too mercurial a temperament to be hedged about, at that stage of his career, by the discipline of mastering a new technique. The Davies's were sadly disappointed: John was an artist they much admired, and several examples of his work were in their art collection.

The reason for Maynard's visit to London was that he was attending an exhibition organised by the First Edition Club at which the Gregynog Press books were having their first major public showing. Captain Abbey was prevailed upon to lend his levant bound copies for the occasion. The books were well received, and Maynard returned to Gregynog content on this score. Nevertheless, clouds were gathering on the horizon.

One thing troubling Maynard was a recent move on the part of the Press Board to amalgamate the interests of the Press with those of the *Welsh Outlook*[4] and to form a joint limited company. The main reasons for his antipathy were that, in his opinion, the two concerns had little or nothing in common and that he feared a further move to undermine the interests of the Press staff. He sought reassurance that, as controller, he would have the power to intercede on an employee's behalf, should there be any conflict between employers and employed, and requested that formal contracts should be drawn up for all members of his staff. His fears were aired in a letter which he wrote to T. J. on 21 May,

> 'As I see it, the interests of the Press staff are identical with those of the Misses Davies. Any attempt for petty advantage by either party must react in an unhappy way. I have tried hard to foster and maintain a spirit of adventure and enthusiasm in each member of the staff. It will be hard indeed if outside influences are allowed to undermine my work.'

One may sympathise with Maynard for, anxious though he was to play his part in securing the welfare of the craftsmen and women under his care, he was no trades unionist at heart, and, as his later letters demonstrate, he desired above all to be left in peace to concentrate on his own work. Though the proposed merger never took place, Maynard's fears remained unabated, and this episode marks the beginning of a growing obsession which increasingly tormented him, and which was to culminate in his resignation in 1930.

In mid-June, Maynard paid a visit to Eric Gill, then living and working at Capel-y-Ffin, in the Black Mountains of Brecon. The interlude provided the controller with temporary relaxation from his problems. Opportunities for mixing with other artists, a means of mental stimulus, refreshment and inspiration, were rarely afforded. This was the great disadvantage of working in the isolation of rural Montgomeryshire, as Maynard and others, who followed him, were to discover. There is a Welsh saying, 'Rhaid wrth haearn i hogi haearn', (One needs iron to sharpen iron), and certainly creative artists working at Gregynog were underprivileged in this respect. Beautiful and peaceful as Gregynog was, it afforded little in the way of inspiration apart from its solitude. Gill promised to come to Gregynog in the autumn, but the visit never took place.

Maynard and Bray were beginning to feel the strain of being solely responsible for all the illustrative work at the Press. Gill had awakened Maynard's interest in the work of David Jones,[5] as a result of which he suggested to T. J. that Jones might be invited to provide a wood-engraved title-page for *Llyfr y Pregeth-wr*, then in preparation:

> 'It is exceedingly difficult for Bray and myself, or perhaps even impossible for us, to undertake successfully the decoration of every type of book; the occasional employment of artists outside would, I think, sweeten the Gregynog cake.'[6]

An exhibition of paintings and sculpture from the Davies's collection, together with eight paintings lent by Hugh Blaker and a few private press publications, which included the Kelmscott *Chaucer* and the Doves *Bible*, was held at Gregynog from June to August, in aid of the Montgomeryshire Hospitals Appeal Fund. For this the Press printed an eight-page hand-list which stated that 'in accordance with the wishes of the Misses Davies, the total proceeds from the sale of Gregynog Press books during the exhibition will be handed over to the funds of the [appeal].'

On 24 June, the printing of *Selected Poems of Edward Thomas* was completed, but not before last minute reprintings were made of the title-page and colophon. When first printed, the title had read *by* Edward Thomas which, as Edward Garnett pointed out, was incorrect, since he and not the poet had made the selection.[7] A pre-publication copy was sent to J. G. Wilson, the manager of Bumpus's London bookshop, as was customary, since he not only placed a substantial order for each

Gregynog title but also acted as unofficial champion of the Press. The copy was bound in forrel, a kind of pseudo vellum, which Wilson considered detracted from the book's appeal, since the binding would be likely to soil easily. Acting upon his advice, the ordinary edition was eventually bound in yellow buckram. This was the first of the books to be printed on Japanese vellum, an odd choice in view of the fact that the book contains only one small wood engraving.

Edward Garnett was later commissioned to write a prospectus note for the book. A letter from Maynard to Garnett, enclosing press notices, throws an interesting light on the unobtrusive way in which the Davies's liked to conduct their affairs. Referring to the reviews, he says,

> 'I ask you to [quote] discreetly, as the owners of the Press very carefully throttle even legitimate self-advertisement!'[8]

This same prospectus was to be the cause of some embarrassment to Maynard, for, the next day, he was writing to Garnett again to confess to a misprint in the specimen page which had been printed to accompany Garnett's note. In the poem entitled *Words*, appeared the opening, 'Out of us all', instead of, 'But of us all'. Maynard asked Garnett whether it could remain unaltered, suggesting that he might print an erratum slip, as the reprinting of the sheet would mean a month's delay. Worse still, he opined that the erroneous line 'read equally well and perhaps more musically'. Deeply shocked, Garnett replied by return, saying that an erratum slip at least was necessary and continued,

> '. . . if one once began letting typographical errors stand because "they read equally well and perhaps more musically", where in heaven would the process stop?'[9]

A chastened Maynard replied that he would reprint the page without delay!

The book was published on 8 August. It was in no way a companion volume to the previous Thomas book, even though it shared the same type face. The text area of the pages is ruled faintly in blue, a feature which T. J. was not sure that he liked at first, but the general effect is harmonious. The wood engraved initials, printed in red, are large, gracefully designed shadow letters which add considerable interest to the page layout, despite the fact that Bernard Newdigate, in his review of the book, suggested that the initials and versals would have looked better had they been supported by lines of type directly beneath.[10] There is an agreeable title-page which bears a wood engraved

Opposite: Title-page to The Life of Saint David, *printed in blue and black. (Inner and outer margins reduced.)*

THE·LIFE·OF SAINT·DAVID

MCM· XXVII
THE GREGYNOG PRESS

✝

floral spray, supported by the poet's initials, enclosed in a double-rule octagonal frame, printed in pale blue. Twenty-five copies were bound in blue levant, to a design by Maynard, in which a gilt foliated pattern is arranged perpendicularly in interlacing squares, connected to the edges by gilt ornament. For this the leaf sprays were cut as a single tool, to save Fisher time in the tooling. The book was an immediate success, all copies being subscribed for within one month of publication.

Sales were, by this time, fairly well assured, though few subscriptions came from within Wales. The books were offered to the retail trade at twenty-five per cent discount on orders of less than twenty copies. In August 1927, Wilson of Bumpus wrote to the Press Board complaining that a rival stockholding bookseller was offering the same books to the trade at one third discount. As a result, T. J. wrote a polite but firm letter of rebuke to the company concerned who offered to conform to practice in future.

During these summer months, the Press staff were busy preparing the next two books but, nevertheless, found time for printing a four-page list of the works published to date, which includes mention of the two forthcoming volumes. They also printed a four-page programme for a concert of vocal and chamber music held on 11 July, during the annual conference of the National Council of Music.

The question of how many review copies of books should be sent out was debated in the autumn by the Press Board. Publicity was desirable, yet, in view of the small size of editions published, the Board could not afford to be over generous in this respect. T. J. sought the advice of Stanley Morison. His letter brought forth a characteristic Morisonian reply on 18 October, to the effect that he was opposed to the gratuitous distribution of books for review and considered that reviewing journals should purchase copies. This opinion was not a great deal of help, as obviously the major reviewing journals would not be prepared to purchase copies. The Press Board eventually came down in favour of minimum distribution, and the number of copies sent out seems to have been decided on an ad hoc basis, as titles were published.

During 1927, experiments were made in printing on vellum, but the splendid vellum editions of other private presses found no rival here. Of these and later experiments only a few examples remain.

The edited text for *The Life of Saint David* had been furnished by Ernest Rhys as long ago as February 1925, yet the printing was not completed until 27 September 1927. This slim quarto, the bravest experiment in illustration so far, was the first book to be printed on *damp* hand-made paper. From this time on, almost all books on hand-made paper were so printed. It seems astonishing that Maynard had only recently learnt of the technique through a chance remark made to him by Bernard Newdigate of the Shakespeare Head Press. In writing to C. H. St. John Hornby on 16 November about the book, Maynard commented,

'The book is not quite such a success, from the technical point of view, as some of the earlier things, but it represents our first effort on damped paper, and we encountered considerable difficulty in getting the three printings to register.'

The hand-coloured wood engravings were also a new departure, the colouring being carried out by the girls in the bindery, working under the direction of Horace Bray. These illustrations, apt and charming and delicately and subtly coloured, are perhaps the most successful feature of the book.

Hodgson's name appears for the first time in the colophon, and his masterly touch is already apparent. Poliphilus type was a felicitous choice and accords well with the illustrations. The text is divided up by quill-drawn paragraph marks in red which Bray inserted himself in each copy. The title-page is an admirable example of restrained design in which the large blue wood-engraved letters of the title are finely balanced by the small black wood engraving of the saint and the small blue cross at the foot of the page.

By good fortune, a small brass, depicting Saint David, had been discovered only a few months previously in Hereford Cathedral. Rhys had brought the find to Maynard's attention, suggesting that a reproduction of it might be used in the book. Early in September, Maynard and Bray motored to Hereford where they took a rubbing of the brass, which was later translated into the block used on the title-page.[11]

The binding of the ordinary edition represents the only use of vellum by the Gregynog bindery. Clearly this was not a material with which Fisher was familiar, for the turn-ins show untidily through the endpapers, a fault which could easily have been obviated by the use of an extra sheet of lining material. Scarlet levant morocco was used for the special binding which is handsomely decorated, in gilt on both covers, with a large Celtic cross, enclosed in a rule frame.[12]

The Life of Saint David was published on 24 November, the same day as *Llyfr y Pregeth-wr: the Book of Ecclesiastes*, the printing of which had been completed on 1 October. This book is of interest chiefly as the first to be illustrated by an outside hand, David Jones being responsible for the wood engraved frontispiece and title-page. It is a competent piece of work but dull, and the wood engravings are too heavy, ill-suited to the remainder of the book and to the paper on which they are printed. Poliphilus type here loses something of its dignity: it would have benefited from closer setting, unfortunately not possible due to the syllabication of the Welsh language. The page effect is not improved by the over-large cruciform paragraph marks. Press records state that two copies of the book were printed on vellum and specially bound, though for whom this was done is not recorded.[13] The ordinary binding is in limp blue buckram and the special binding in dark blue levant morocco, tooled with the title in gilt on the front cover.

Maynard judged the time ripe for issuing a printed invitation to subscribers. A broadside was printed, inviting subscriptions for *The Life of Saint David*, *Llyfr y Pregeth-wr*, *Poems by Henry Vaughan* and *Caneuon Ceiriog*, the last two mentioned having proved slow sellers. *The Life of Saint David* is here described as being 'in full forrel binding', but the word 'forrel' is deleted by hand and the word 'vellum' inserted.

Two vastly different Christmas cards were produced in 1927, though both were designed to bear personal greetings from the sisters and their stepmother. The simpler of the two is a horizontal card on which the greeting is surrounded by a border of fleurons. The other is a double fold of Japanese vellum, bearing the greeting, printed in red and black, on the front, which encloses a single sheet of Japanese vellum on which are printed four verses of Ben Jonson's *A Hymn on the Nativity of My Saviour*, printed in black with red initials, and surrounded by a double rule frame in red.

The Press could rest content with the year's achievement: for the first time it had managed to bring out three books within twelve months.

An event occurred towards the end of the year which typifies Gwen Davies's unobtrusive generosity. It came to her ears that a Welshman, whom she knew well and who had, so far, subscribed to all the Press publications, had received no salary for some months past, as his employers were in dire financial straits. He had vowed to keep up his subscriptions, even if it meant giving up smoking. Gwen begged T. J. to join with someone else in giving their mutual acquaintance a copy of *The Life of Saint David* as a Christmas gift, for which she offered to foot the bill. The present was duly given and its true source never revealed.

Engraving by David Jones from Llyfr y Pregeth-wr.

Early in 1928, Maynard began to grow restive. Perhaps this was due to personal malaise, something which, sooner or later, affected all the Gregynog artists. Partly it may have been due to overwork. The range of his responsibilities was growing, and the network of individuals with whom he had to deal became more complex with each succeeding year. The extension of administrative duties began to come between Maynard and his creative work, to which he was able to devote less and less time.

The Welsh Outlook was now owned by the Davies's, and, in an attempt to give the journal a new and brighter image, Maynard, assisted by Dora Herbert Jones, often worked until late in the evening at the Newtown office, helping the printer, Gilbert Phillips, to improve the layout. Also in February, Maynard was called in to advise William Lewis, the Cardiff printer, on the layout of a booklet containing the laws of Hywel Dda which was to be published by the University of Wales Press. For this Maynard drew versals and a title-page heading. Moreover, it seems likely that the cover and title-page were actually printed at Gregynog, though there are no records to prove that this was so.

A month before, Maynard had again become involved in a dispute with the agent of the Gregynog estate who, with the Davies's support, was insisting that all matters concerning staff housing should be discussed with him. Maynard, who had fought consistently for the right to negotiate such matters on behalf of his staff, took umbrage at what he considered an unwarranted attempt to undermine his authority. He informed Gwen Davies that he intended seeking accommodation away from Tregynon and requested an increase in salary to enable him to purchase his own car and typewriter. He threatened resignation, should his request not be granted.

> 'Nobody knows better than yourself how unreservedly I have given of myself in carrying out the work I undertook for you, and I am heartbroken to see the fabric I built with so much care undermined, and my long services counted for nought.'[14]

At T. J.'s behest, W. J. Burdon Evans intervened and managed to calm Maynard, so that his threat was not carried out, but the atmosphere of mutual distrust between agent and employees was to continue throughout the history of the Press.

Earlier in January, the Press Board had a slight brush with Stanley Unwin[15] when they had requested, through the author, permission to reprint without charge Gilbert Murray's translations of the *Plays of Euripides*. Murray's letter brought forth a characteristic Unwin reply:

'The Press may be a labour of love, but the productions are not given away, and I am not quite clear why the owners of the Press should not pay a royalty on such copies as they sell, as other publishers have to do.'[16]

The Board finally agreed to pay Unwins a royalty of fifteen per cent of the published price, and all was amicably resolved.

Controversy arose, later in the spring, over the Welsh spelling of the name of the Press. The form 'yng Ngregynog', used in the colophon to *Llyfr y Pregeth-wr*, had been criticised by W. J. Gruffydd in his review in *Y Llenor*.[17] He claimed that the form 'yng Ngregynnog' was more correct. The matter was hotly debated, and finally a ruling was sought from Sir John Morris-Jones, the Welsh scholar and orthographer, who supported Gruffydd's view. The debate was purely academic, since the true origin of the name is unknown. This issue was based on the assumption that the name was derived from 'cregyn' (abounding in shells), but the case was never satisfactorily proven. The incident gave Maynard an idea. He had always disliked the Press device motif. Here perhaps was the opportunity to change it – to a shell? The shell device was, in fact, not used until after Maynard had left the Press, but doubtless Hughes-Stanton, who designed it, drew his inspiration from Maynard.

In April, the Press published *An Account of the Convincement, Exercises, Services and Travels of that Ancient Servant of the Lord, Richard Davies* . . . which Maynard described as, 'quite the most sober thing we have done'[18] – to which one might add 'or were ever likely to do'. One writer[19] declares that 'it stands as a flawless achievement – no decoration, no second colour except on the title-page – just Baskerville type on good paper'. All of which is undeniably true, but it is a monumentally dull book for all that. The Press Board were perhaps wise to limit the edition to one hundred and seventy-five copies! Nor does the binding, in plain navy blue buckram, render it in any way memorable. *An Account* it is called, and as inspiring as an account book does it in truth appear. The special binding, in dark blue levant, does nothing to whet the appetite either. It is plain except for gilt spine titling and the spine bands tied down with blind ties. The marvel is that Fisher managed to get such a mouthful of a title on to so slim a spine!

The first folio to come from the Press, *The Autobiography of Edward, Lord Herbert of Cherbury* was published in May. The appearance of this book certainly raised the Press from the ranks of the ordinary and marks a new beginning in its fortunes. Publication evoked lyrical reviews from all quarters. Even the reviews in *The Manchester Guardian* and *The*

Observer[20] discussed its physical presentation at some length. *The Observer* review noted,

'. . . the hand-made paper is of a stuff to last forever. The type is a perpetual grace – an old-face, dark and tall like Herbert himself, with refinements accentuated, not obtruded. The italic letter of the introductory pages is something to look at before you read it, and to look at again afterwards. The illustrations are woodcuts like the old folios, but more defined to harmonise with the serried type.'

It would be difficult to find pages where Poliphilus and Blado are shown to better advantage, and the large initial openings, printed in red, are a *tour de force*. If there had been any doubt previously concerning Maynard's typographical skill, he is fully vindicated here. The title-page which is one of classic simplicity, emblazoned with the Herbert arms, hand-coloured in red and blue, alone is a work of art. The nine large wood engravings by Bray add greatly to the period flavour and are quite unlike any of his previous work, particularly in their meticulous detail. The printing of these, on the roughish Batchelor hand-made paper, represents a triumph for Hodgson.

It is a pity that the ordinary binding is so dull. It is in brown buckram, blocked with the arms in gilt on the front cover, and one could wish, for this book, something more imaginative. The special binding is much more appealing. It is in dark brown polished levant morocco, plain except for gilt spine titling and the arms on the front cover which are onlaid in red and blue and blocked with the three lions rampant in palladium.[21]

J. L. Garvin, not content with giving the book an enthusiastic review, went on to praise it extravagantly in a letter to T. J.:

'I tell you, Tom Jones, you add grace to the soul of our land, which for this, among other things, you shall not die, when you do such works, you old Gregynog Caxton!'[22]

The letter provides ample evidence that, in the eyes of one of his friends at least, the Press belonged very much to T. J.

Following this, there was a gap of almost six months before the publication of *Selected Poems of W. H. Davies*, arranged by Edward Garnett. It is a pretty straightforward work, printed in Baskerville on Japanese vellum with a simple yet beautiful title-page where the most significant feature is a new wood engraved device, printed in black and yellow. This is nicely complemented by the dark mass of the frontis-

Opposite: Page from The Autobiography of Edward, Lord Herbert of Cherbury, *with engraving by Horace Bray.* (*Reduced.*)

a table for me, and his pages to attend me, the chief of whom was Monsieur de Mennon, who proving to be one of the best horsemen in France, keeps now an academy in Paris ; and here I shall recount a little passage betwixt him and his master, that the inclination of the French at that time may appear ; there being scarce any man thought worth the looking on, that had not killed some other in duel.

Mennon [? Menou] desiring to marry a niece of Monsieur Disancourt, who it was thought should be his heir, was thus answered by him : "Friend, it is not time yet to marry ; I will tell you what you must do. If you will be a brave man, you must first kill in single combat two or three men, then afterwards marry and engender two or three children, so the world will neither have got nor lost by you ;" of which strange counsel Disancourt was no otherwise the author than as he had been an example at least of the former part ; it being his fortune to have fought three or four brave duels in his time.

And now, as every morning I mounted the great horse, so in the afternoon I many times went a-hunting, the manner of which was this. The Duke of Montmorency having given orders to the tenants of the town of Merlou, and some villages adjoining, to attend me when I went a-hunting, they, upon my summons, usually repaired to those woods where I intended to find my game, with drums and muskets, to the number of sixty or eighty, and sometimes one hundred or more persons ; they entering the wood on that side with that noise, piece portrait of the poet, engraved by Maynard.[23] The setting of the verses appears to be pitched a little too much to the left of the page, especially in poems like *School's Out* where the lines are particularly short, but the balance is redressed somewhat by a pale red page ruling.

The ordinary binding is one of the most interesting to date, having a black buckram spine and fore-edge strip, framing sides of a Cockerell combed-marble paper. The paper is found in variants of black on buff or blue on green. Maynard was responsible for the design of the special binding which is executed in dark sage green polished levant bearing six squares of foliate gilt decoration, bordered top and bottom by a four-line gilt rule which is carried across both covers and spine and continued on the turn-ins.

Engraving by Horace Bray from The Misfortunes of Elphin.

Late in November, appeared Thomas Love Peacock's *The Misfortunes of Elphin* which is the first Gregynog book printed wholly in monochrome. The text, set in Baskerville, is enlivened by twenty-one small wood engravings by Bray which form chapter heads and tail-pieces. These give a new slant on the artist's versatility, being simple designs which incorporate squarish stylised figures. The overall impression is one of demure drollery, entirely suited to Peacock's style. The title-page is one of simple, classic beauty. Indeed, the book has a satisfying unity which has rendered it one of the more keenly collected items.[24]

The ordinary binding is again an attractive one, having a purple buckram spine, with titling in gilt, and patterned cloth sides in shades of purply-blue and black. Bray designed the rich red polished levant cover for the special binding, on the front of which appears a small gilt-blocked design of two hand-held goblets.

1928 was a bumper year for the Press, five books in all being published. The last of these, Sir John Morris-Jones's translation into Welsh of *The Rubaiyat of Omar Khayyam*, was so long in the making, due to protracted exchanges between translator and controller, that it deserves a chapter to itself.

The Press was also kept busy with more ephemeral printing which included two prospectuses – one for *The Autobiography of Edward, Lord Herbert of Cherbury* and a composite one for *Penillion Omar Khayyâm*, *Selected Poems of W. H. Davies* and *The Misfortunes of Elphin*, a list of the first ten books, an appeal for the Welsh School of Social Service on Distress in the Mining Areas of Wales and two Christmas cards. The larger of the two cards is an ambitious affair, consisting of a leaf bearing passages from *The Pilgrim's Progress*, printed in Poliphilus type, in red and black, which is enclosed in blue paper covers, bearing the general greeting 'from Gregynog'. The other is a small card, bearing greetings from the Misses Davies and Mrs Edward Davies, again printed in Poliphilus, with a large wood engraved A in the centre, in red.[25]

With the increase in the number of books produced, the bindery staff had been augmented during the year. Fisher was fully occupied in producing the special bindings, and it was becoming increasingly difficult to ward off requests for full leather bindings made by the Davies's friends. T. J. was tactfully admonished by Maynard, on one occasion, for having rashly promised that Fisher would bind a copy of George Bernard Shaw's *The Intelligent Woman's Guide to Socialism* for the author's wife, Charlotte. Maynard pointed out that the Secretary was forever having to refuse such commissions.[26] (The Shaws were friends of the sisters and visited Gregynog on several occasions.)

Notable visitors to the Hall during the year included the Prime Minister, Stanley Baldwin, who was brought by T. J., following a visit to David Davies at Llandinam, and Leonard, later Sir Leonard, Woolley, the archeologist, and his wife, Catherine. Adrian Boult, accompanied by the soprano Elsie Suddaby, made the first of many visits to Gregynog in May, following an appearance at the Montgomeryshire County Music Festival.[27] From 1928 onwards, the visitors' books include many celebrated names.

NOTES

1. The Abbey collection of bindings was sold at auction by Sotheby's 1965–1967. That portion including the complete collection of Gregynog special bindings was sold on 6 November 1967.
2. The printing was carried out in a small back room, first in Paddington and later in Shepherd's Bush. Manning Pike was editor and general manager of the project.
3. The book was eventually published by the Press in 1933, illustrated by Agnes Miller Parker.
4. *The Welsh Outlook*, 1914–1933. A monthly journal of Welsh social progress, originally financed by David Davies, which was edited by T. J. for the first three years. Until 1920 it was printed in Cardiff, after which it was published under the patronage of Gwen and Daisy Davies and printed in Newtown by the Montgomeryshire Express Company Ltd.
5. David Jones, 1895–1975, artist and author.
6. Letter from Maynard to T. J., dated 14 June 1927.
7. On the half-title of the book, the error stands.
8. Letter from Maynard to Garnett, dated 8 July 1927.
9. Letter from Garnett to Maynard, dated 11 July 1927.
10. Review by Bernard Newdigate in *The London Mercury*, Vol XVI, September 1927.
11. Maynard recorded that the brass was only eight inches high and very worn.
12. The design is by Horace Bray.
13. Numbers 203 and 249, according to Thomas Jones's bibliography.
14. Letter from Maynard to Gwen Davies, dated 4 January 1928.
15. Sir Stanley Unwin, 1884–1968, Chairman of Allen & Unwin Ltd.
16. Letter from Stanley Unwin to Gilbert Murray, dated 14 January 1928.
17. William John Gruffydd, 1881–1951, Professor of Celtic Languages, University College Cardiff, 1918–1946. Editor of *Y Llenor*, 1922–1951. The review appears in *Y Llenor*, Winter, 1927.
18. Letter from Maynard to the editor of *The Welsh Outlook*, dated 26 March 1928.
19. Colin Franklin in *The Private Presses*. Studio Vista, 1969.
20. Reviewed in *The Manchester Guardian*, 25 June 1928. Review signed HBC (presumed to be Henry Buckley Charlton, the Shakespearian scholar). Reviewed in *The Observer*, 10 June 1928, by J. L. Garvin.
21. The palladium areas were originally tooled in silver which, of course, tarnished fairly rapidly. When this was discovered, all copies were recalled for re-tooling.
22. Letter from J. L. Garvin to T. J., dated 28 June 1928.
23. In order to achieve the particular matt effect Maynard desired for the portrait, Hodgson mixed white spirit with the printing ink.
24. Both E. M. Forster and Michael Sadleir were asked to furnish an introduction for the book which each in turn declined, claiming that he had little knowledge of Peacock.
25. There is also a version of this card printed on vellum.
26. Letter from Maynard to T. J., dated 18 July 1928.
27. An annual event, held in Newtown, which had been started with generous financial support from the Davies sisters.

7
Penillion Omar Khayyâm

SIR JOHN MORRIS-JONES'S TRANSLATION INTO Welsh of the *Rubaiyat of Omar Khayyam*, scheduled to appear in 1927, was not published until December 1928. The road to publication was a long and thorny one for Maynard who had to contend with much well meant interference from the translator. Sir John had forthright views on typography, layout and illustration, concerning which he was surprisingly knowledgeable, views he was prepared to air at length. This was Maynard's first real battle with an author. The first stages of the contest he undoubtedly found stimulating, if exacting, but, as time went on, and publication was delayed again and again, his patience wore a little thin.

In May 1927, the Press Board had approached Sir John Morris-Jones[1] with a request to be allowed to publish his translation which had first appeared in 1907, in his *Caniadau*. The author consented with alacrity and waived all claim to any royalty payment. T. J. had conceived the idea of publishing the work in time for the National Eisteddfod to be held in August at Holyhead, but he had reckoned without Sir John.

In order to expedite publication, the Press planned to prepare proofs from the original edition for correction by the author, but Sir John was shocked by such haste. In June, he wrote to say that he had found many more errors in the original printing than he had previously been aware of. He went on to suggest that the book might be decorated with a few Persian ornaments and some illustrations of Omar on the lines of those conceived, some years earlier, by Gilbert James. On 8 June, Maynard entered into correspondence with Sir John when he wrote outlining his plans for the book concerning page size, typeface and other matters.

'It seems to me that your translation of Omar Khayyam should either be treated in a very simple way – relying entirely on the spacing and proportion of the type – or that some abstract decoration, possibly in colour, based on Persian ornament might be introduced.'

Confessing unfamiliarity with Gilbert James's work, he continued,

> 'I do not care for anything which suggests realism in decorations intended for poetry of an imaginative or philosophic nature.'

Maynard's choice of Caslon Old Face and proposed page format were applauded by the author in his reply, but concerning Maynard's views on the decoration of the book he was not so happy:

> 'I am not sure that Persian ornament would go very well with Caslon type; and I feel that the illustrations such as I had in mind would convey the oriental atmosphere better than any amount of oriental ornament.'[2]

Faced with this opposition, Maynard was forced to capitulate and to agree to design some wood engravings on the lines suggested, but, alas, any hope of publishing the book in time for the Eisteddfod had to be abandoned, and there matters rested until March of the following year.

Late in March 1928, Maynard wrote to Sir John enclosing galley proofs of the text which had been set from the original version. He pointed out that he had altered the layout of the lines so that they were all ranged left, whereas in the previous version each line, except for the third in each quatrain, was indented. Maynard maintained that the new layout improved the appearance of the page. Where a line of asterisks had previously marked the division of the poem into periods or days, Maynard proposed introducing a coloured initial at the opening of stanzas 16, 28, 41 and 73. He informed the author that he was working on the illustrations but that, as his knowledge of Omar was based entirely on the Fitzgerald translation where the grouping of events differs, he was experiencing some difficulty.

Sir John was almost a month returning the proofs. A long letter accompanied them, dealing with the points made by Maynard and raising others. He agreed to the new line setting and to the use of initials to mark the beginning of sections but took Maynard up on the use of equal spacing after punctuation, claiming that increased spacing after a full stop was preferable, especially where a pause was indicated in mid-line. In his reply, Maynard concurred with Sir John. He went on to say that he had now had an opportunity to see the illustrations of Gilbert James and could not agree to basing his illustrations upon them, as he found them altogether too realistic for their purpose. He promised to send Sir John proofs of his own illustrations in due course.

As in the previous year, there was pressure from the Press Board to get the book out in time for the National Eisteddfod. Sir John was urged to pass galley proofs so that the pages could be made up in readiness for the illustrations to be added,[3] but he was not to be rushed and refused to pass anything except *page* proofs for publication.

'I am not keen on getting the book out before the Eisteddfod. I am much keener on getting it out perfect.'[4]

His eagle eye also picked out footnotes which had been set in the same type size as the text. To this he quite properly objected and suggested a smaller size.

By late June, Maynard sent the author page proofs, together with proofs of three wood engravings and tracings of three other designs, explaining that, in his view, the designs should not only bear some reference to the text but that, more particularly, they should decorate the book and should collectively create a total atmosphere.

If Maynard hoped for blanket approval, then he was certainly disappointed. Sir John replied to the effect that he liked the designs and thought Maynard successful in having achieved atmosphere. Nevertheless, he wanted to change details in every one of them in order that they should conform *exactly* to the wording of the text! With great forbearance, Maynard agreed to minor changes where possible and attempted a defence of the design as it stood, in cases where it would have been well-nigh impossible to alter it. Sir John, however, was not mollified and wrote again supporting his views. Accuracy of detail was obviously all-imporant to him.

'What I feel is that, in a book such as ours, specially illustrated, there ought to be perfect harmony between the matter and the decoration. This you have achieved triumphantly, I think, in the style and techniques of the blocks, but I think also that it is worth taking a little trouble to try to obtain perfect harmony in detail, so that the pictures fit the verses like a glove.'[5]

Sir John laid stress on the essential differences between his rendering and that of Fitzgerald and implied that he would appreciate the opportunity to comment on the designs before they were transferred to the wood. Crushed, Maynard had to agree to recut at least one of the blocks, and, yet again, the idea of producing the book in time for the Eisteddfod had to be abandoned.

There followed a barrage of correspondence concerning various aspects of the illustrations. In July alone, Sir John wrote seven lengthy letters dealing with all manner of specific detail and, again and again, referring Maynard to finer points of the text. These covered such things as a play upon the word *gūr* which has variant meanings in Persian of *wild ass* and *grave*, the state of neglect of Bahrām's Hall, the omission of a mouth on a face, suggestions as to how to incorporate two uncontemporary events in one picture, suggestions for the portrayal of simile, which included a lengthy and didactic explanation of the difference between simile and metaphor, and a suggestion as to the precise form a newly conceived drawing might take.

Maynard was beginning to grow restive under such treatment and, on 16 July, penned a long letter to Sir John, pointing to the crux of the dissension between them:

'. . . it is actually a sharp conflict of views as to the province of the artist or decorator. Is he called in merely to give expression to concrete verbal images, or is he allowed to give expression to what he feels about his text and in his own way? . . . my idea in these Omar decorations is not especially to illustrate specific things in the text, but rather to summarise the general impressions I have of the book, and to do it in such a way that the right atmosphere is created. Above all to decorate. To decorate is the very first consideration in the architecture of the page, and if he fails in this, the artist fails all the way along the line, and his work, no matter how true and how relevant, is wasted. You may be able to see your visual image in a decorative way, but it does not necessarily follow that I can. Perhaps William Blake, for this reason, was the happiest of men.'

In his reply of 20 July, Sir John aired his divergent views on art in general and on book art in particular. Difference of opinion had arisen over one drawing in particular.[6]

'It is very distasteful for me to criticise the drawing of which you seem to have a high opinion, but I must ask you to consider very seriously whether it does not transgress a fundamental law of all art. Aristotle's doctrine that a poem . . . should be one organic whole, in which the parts are interdependent . . . is surely held now to be true not only of poems but of every work of art. Now this drawing has no unity; it is the putting together of unrelated and incoherent parts. I do not think you can rely on the quatrain to lend it a unity which it does not itself possess.'

A few suggestions as to how the illustration could be tackled followed and then he went on to say,

'After all I have said, I wish you to understand that I do not desire to oppose your considered judgment. The responsibility for the decoration is yours, not mine.'

Sir John concluded by opining that all book decoration should illustrate the matter and culled examples from the history of book illustration to prove his point. That examples could be found equally to disprove his theory seems to have escaped him!

Maynard, unabashed, countered the onslaught by claiming that artistic unity depends solely on the correct relationships of design and shapes to which the literary significance of the various parts might be unrelated. He argued that design only loosely connected with the text allows the reader to use his own interpretive powers. Nevertheless, he did concede that there were circumstances where an illustration needed to preserve the literary unities and offered to make another attempt at the drawing in question. This task he was happily spared as, due to a necessary rearrangement of the text pages, it was found that the quatrain could not fall opposite the space left for the illustration.

Attention was now focussed on the need to illustrate Omar's preoccupation with astrology and philosophy. Sir John once again entered the arena with much enthusiasm and attention to detail. He made a rough sketch of a fifteenth century Italian woodcut depicting an astrologer and made suggestions as to how this might be amended and used as a basis. His descriptive outline revealed a deep knowledge, not only of Persian social history and costume, but of astrology too. This was only a beginning: more was to follow.

Meanwhile, Maynard had designed a new device for the title-page and courteously invited Sir John's opinion of it. Praise from that quarter was never unqualified, and, while the author approved the title-page in general, he considered the imprint too weak and too insignificant. (The new form GWASG Y GREGYNNOG, incidentally, was applauded.) He further requested that the square brackets which Maynard had used in the introduction be replaced by curved ones and enclosed two rough drawings of the planetary system as it might have been drawn by Omar. These were accompanied by precise details as to how they might be translated into a satisfactory illustration.

By 6 October, Maynard intimated to Sir John that the author's failure to pass proofs of the introductory matter was causing a hold up in the Press, as Hodgson was running short of work. The proofs came by return, but the accompanying letter raised further questions concerning a word on the title-page, about the use of brackets, about one of the blocks and about the setting of a rule beneath the colophon. There was to be no let-up for the long-suffering controller. All these points he answered or adjusted, as seemed possible to him, but the matter of the astrological illustration was still not resolved. His proposed design brought forth several protests from Sir John: Saturn, the all-important sign, referred to in quatrain 48, was obscured on the planetary chart, the sign of Mars was incorrectly drawn, the compasses held by Omar were incorrectly placed, and Sir John took exception to Omar's beard. This, he said, reminded him of Calvin, whereas he favoured a more patriarchal style. Omar, he claimed, was too tall and too youthful. Maynard did his best to accommodate all the author's amendments when redrawing the design, but he stood firm in defence of another illustration for which the block was already engraved.[7]

'In any case, for the reasons stated . . . , namely expense, it must stand; for the book has already cost far more than the rather arbitrary price which the owners fixed to make it available for the Welsh public . . .'[8]

For once, the author had to climb down.

'As you say, with regard to the woodcut of the hunter, that the question of cost is decisive, I, of course, have not a word more to say.'[9]

By mid-November, Sir John was still not happy with the design for the astrological block. The planets, he complained, were incorrectly drawn, and he sent notes for their correction. Omar was still too tall, too young and too slender, and he even found fault with the table, enquiring why it need be so low. Though Maynard had redrawn Omar's beard, the author was still not satisfied.

'I do not think that a beard like that of a goat is at all correct for an old Persian. The Persian beard is full: it spreads out under the ears (it comes high up on the cheeks) and ends either square or with the angles rounded off. Perhaps it might end in an obtuse angle.'[10]

On 28 November, Maynard wrote to T. J.,

'Thank heaven the Omar is completed! I began to get very weary of the wood-cutting – working against time & right through the weekends. I pray Sir John Morris-Jones will be pleased – somehow I ha' ma doots . . .'

Maynard and Sir John never met. One wonders why on earth not. Much time, energy and ink could have been saved on both sides!

The book was published on 7 December, in time for the Christmas

Opposite: Page from Penillion Omar Khayyâm, *with engraving by Robert Maynard.*

L

I'r Cyngor cudd nid oes a ddyry lam;
Ni chemir dros y terfyn hanner cam;
Oddi wrth y disgybl at yr athro trof,
A dirym megis minnau pob mab mam.

LI

Tynged â'i gordd a'th yrr fel pêl ar ffo
I ddeau ac i aswy yn dy dro:
Y Gŵr a'th fwriodd i'r blin heldrin hwn,
Efô a ŵyr, Efô a ŵyr, Efô.

market. The author received four copies and wrote immediately to Maynard to express his pleasure,

'I thank you very sincerely for all the trouble you have taken with the book.'

Even this letter, however, contains advice. He did not like the look of the ink which had been used for numbering copies, as he thought it would quickly fade and recommended,

'Blackwood's Old English Record Writing Ink which is used in all Government records. I have copies of MSS written in it before 1900, and the ink is to-day as black as the best printing ink.'[11]

Ten days later, came the final bombshell.

'I am sorry to say that the "really damnable" error which you feared might have crept in, has actually occurred.'[12]

This referred to the omission of an initial F at the opening of quatrain 41. He was certain that he had marked it in proof. The implication was that the fault lay with the compositor. Sir John enquired whether there was any possibility of reprinting the page in the twenty-five special copies yet unbound. Naturally, Maynard was dismayed but wrote defending the Press, after checking the proofs, where it was plain to see that the author had not marked the omission. He promised to reprint pages for the special copies. Sir John, obdurate in argument, proved magnanimous in defeat.

'Peccavi! The fault is mine. I had been so anxious to make a separation of the 5 sections clear that I could hardly believe such a lapse as this possible if I had not the evidence of my own eyes. Please convey my sincere apology to the compositor for laying the blame on him. His share of it is slight compared with mine . . . I am extremely glad to hear that you consider you must reprint the . . . pages . . . for the Levant copies. I thank you most sincerely. It will be a great consolation to me to possess one copy which will be an unmixed delight.'[13]

Alas, Sir John was not to experience this delight, for he died the following April.[14]

After all Maynard's struggles with the illustrations, the result appears not to have been recognisably his, for when Edward Garnett received his copy, he wrote enquiring the name of the illustrator!

To the onlooker, long after the event, the result is pleasing enough, despite the fact that the type used for the introduction and the notes is too small for easy reading. The title-page is particularly attractive. The ordinary binding, by comparison, is singularly unattractive, being covered in white buckram sides with yellow buckram spine and fore-edge strip. The white buckram is not smooth surfaced, with the result that most copies have become very soiled with the passage of time, while the onlaying of one buckram on another is surprisingly 'lumpy'. The special binding, in light blue levant morocco, is entirely plain, except for gilt spine titling.

NOTES

1. Sir John Morris-Jones, 1864–1929. Scholar, poet, critic, Professor of Welsh at University College Bangor and expert on Welsh orthography.
2. Letter from Sir John Morris-Jones to Maynard, dated 10 June 1927.
3. It should be noted that illustrations were invariably printed separately from the text at Gregynog.
4. Letter from Sir John Morris-Jones to Maynard, dated 17 June 1928.
5. Letter from Sir John Morris-Jones to Maynard, dated 4 July 1928.
6. The illustration relates to quatrain 52, concerning fate and the chessboard.
7. This refers to the illustration of Bahrām at the beginning of the notes.
8. Letter from Maynard to Sir John Morris-Jones, dated 27 October, 1928.
9. Letter from Sir John Morris-Jones to Maynard, dated 30 October 1928.
10. Letter from Sir John Morris-Jones to Maynard, dated 15 November 1928.
11. Letter from Sir John Morris-Jones to Maynard, dated 8 December 1928.
12. Letter from Sir John Morris-Jones to Maynard, dated 18 December 1928.
13. Letter from Sir John Morris-Jones to Maynard, dated 20 December 1928.
14. In making the briefest extracts from this fascinating battle of wills, I have chosen to highlight only those portions which relate chiefly to the illustrations, but the correspondence contains infinitely more minute points concerning punctuation, spacing, type size and weight, etc., etc. For a full account the reader is referred to an article in *The National Library of Wales Journal*, Vol. XVII, No. 1, Summer, 1971, where David Jenkins has reproduced the correspondence.

8

Psalmau Dafydd

THE YEAR NINETEEN-TWENTY-EIGHT ENDED ON a sad note, for, on 21 December, occurred an event which shattered the community at Gregynog. Elphin Lloyd Jones, T. J.'s youngest child, was killed by a car near his home in Hampstead, as he went to post some letters. A clever, quick-witted child with a wisdom beyond his years, Elphin was the treasure of his father's heart. T. J. was utterly broken. To Gwen Davies he wrote later in the day,

'The lovely life is ended – brutally, stupidly, and the ability and the fun and the old-fashioned sense "finished" . . . I feel as though I had no feeling.'

Not surprisingly, following the tragedy which had hit T. J.'s family, the first piece to come from the Press in the new year struck a sombre note. To the post-Christmas party of friends gathered at the Hall, from 5 to 7 January, came a message from T. J. and his wife, René, which the Press printed on Japanese vellum. The message read:

'Give our affectionate greetings to the group, and should they wonder about us, say we feel freshly dedicated to helpfulness of those whose sufferings and losses are more hard to bear than our own, with fewer to share them – Ever, T-J.'

Comparatively, 1929 was not a very eventful year for the Press, though much was in active preparation for future publication. One work only was published during the year, and that not until October, but this book, the *Psalmau Dafydd*, taken from William Morgan's Welsh Bible, and edited by Ifor Williams, was arguably the most ambitious Welsh language publication to come from the Press. It is also the most highly decorated book of all. The colophon records that the printing was completed on 10 April, but it seems likely that publication was deliberately postponed to catch the Christmas market.

The verse text, set in Poliphilus type, is exquisitely printed on Batchelor hammer and anvil paper and has elaborate foliated openings and decorated wood-engraved initials, in red and blue, designed and cut

Opposite: Title-page to Psalmau Dafydd, *printed in red and black, engraved by Horace Bray.* (*Margins reduced.*)

PSALMAU DAFYDD

YN OL

WILLIAM MORGAN

1588

GWASG GREGYNNOG

1929

by Bray. Even the colophon is enclosed in vine leaf tendrils, but most elaborate of all is the wood-engraved title-page. It is, above all, a visual book, and it is perhaps not surprising that the edition sold out rapidly, even to those who could not read it! Yet it is an acquired taste, for to some eyes it will inevitably appear a little overdressed.

The artist was also responsible for the design of the paper used in the binding which has a red niger morocco spine, with five raised bands outlined in blind, to tone with the colours of the paper, and buckram-tipped corners. The special binding is in plain blue levant morocco, having gilt spine titling and turn-ins richly tooled with fine gilt rules.

More ephemeral printings included libretti for Bach's *Jesu, Priceless Treasure* and Walford Davies's setting of Richard Baxter's *Lord it Belongs Not to My Care* (both of which works were performed at a Gregynog concert on 26 April), two prospectuses announcing future publications, one issued in May and the other in September, and a separate bilingual prospectus for the *Psalmau Dafydd*. In 1929, the Press undertook, for the first time, the printing of the certificate presented at the annual national eisteddfod of the Urdd Gobaith Cymru (the Welsh League of Youth). This is a well-designed item, printed on hand-made paper, which the Press continued to print annually until 1939. The layout, over the years, changed only slightly, though a variety of type-faces was used.

The 1929 Christmas card was on a more modest scale than those of the two previous years. A cover, bearing a greeting from the Davies sisters and their step-mother, encloses a two-page insert on which are printed a wood engraving and three lines from Herrick:

> 'Dark and dull night, fly hence away
> And give the honour to this day
> That sees December turned to May.'

Apart from the *Psalmau Dafydd*, the most interesting item to appear in 1929 is *Elphin Lloyd Jones: a Memoir*, printed in March for private circulation. It is a charming booklet of twenty-four pages, set in Baskerville type and printed on Japanese vellum, which contains a reproduction of one of Elphin's school essays, separate memoirs by his parents and three engravings by Maynard. It is bound in grey-blue paper-covered boards, decorated on the upper cover with a circular wood-engraved device, bearing the initials ELJ in intertwined ropework, attached to an outer circle which bears the legend BARRY 22 JANUARY 1916:

Opposite: Page from Psalmau Dafydd (*printed in red and black*), *with engraving by Horace Bray.* (*Slightly reduced.*)

Psalm I. *Beatus vir qui non abiit.*

GWYN EI FYD y gŵr ni rodiodd yng-hyngor yr annuwolion,
Ac ni safodd yn ffordd pechaduriaid,
Ac nid eisteddodd yn eisteddfa gwatwar-wŷr:
Onid bod ei ewyllys ef yng-hyfraith yr Arglwydd:
A mefyrio o honaw yn ei gyfraith ef ddydd a nôs.

Canys efe a fydd fel prenn wedi ei blannu ar lann dyfroedd,
Yr hwn a rydd ei ffrwyth yn ei brŷd:
A'i ddalen ni ŵywa,
A pha beth bynnac a wnêl, efe a lwydda.
Nid felly y bydd yr annuwiol,
Onid fel mân us yr hwn a chwâl y gwynt ymmaith
oddi ar wyneb y ddaiar.

Am hynny yr annuwolion ni safant yn y farn,
Na'r pechaduriaid yng-hynnulleidfa y rhai cyfiawn.
Canys yr Arglwydd a edwyn ffordd y rhai cyfiawn,
A ffordd yr annuwolion a ddifethir.

Psalm II. *Quare fremuerunt.*

PAHAM y terfysca y cenhedloedd?
Ac y bwriâda y bobloedd yn ofer?
Y mae brenhînoedd y ddaiar yn codi i fynu,
A'r pennaethiaid yn ymgynghori yng-hyd,
Yn erbyn yr Arglwydd, ac yn erbyn ei Grist ef.
Drylliwn eu rhwymau hwy:
A thaflwn eu rheffynnau oddi wrthym.

HAMPSTEAD 21 DECEMBER 1928. The engraving later appeared on a book-plate designed for use in books given, in the boy's memory, to Coleg Harlech, an institution in which he had evinced particular interest.[1]

In June, the Director of Education for Montgomeryshire wrote to the Press Board enquiring about the possibility of purchasing copies of the books for the County Library service. T. J.'s reply, on behalf of the Board, that,

> 'The books printed at the Gregynog Press are not suitable for circulation by a County Library,'[2]

says much for the low esteem in which public libraries of the time were held, even by one so enlightened as T. J.

A testimony of the growing fame of the Press is a letter received at Gregynog in July from a Mr Koh Hirai of Tokyo who was, so he claimed, the owner of the first private press in Japan. The writer promised to review *Poems by Henry Vaughan* in various Japanese periodicals and also offered to send the Press samples of Japanese papers should these be of any interest. Mr Koh Hirai specifically requested that any mail sent to him should be marked 'via Siberia', as this would save time![3]

T. J. was made a Companion of Honour in 1929 in recognition of his services to the Cabinet, an event which delighted his many friends and acquaintances.

NOTES

1. A residential college for adult education, founded at Harlech in 1927, of which T. J. was one of the founders.
2. Letter from T. J. to Llewellyn Phillips, dated 19 June 1929.
3. Letter from Koh Hirai addressed to the Press, dated 9 July 1929.

9

Maynard and Bray Resign

BY THE BEGINNING OF 1930, MAYNARD HAD BEEN at Gregynog seven and a half years and in the Davies's employ for nine. Apart from the state of friction which existed between him and the estate manager, his relationships with his employers and with those under his charge had always appeared outwardly calm and unruffled. Then some action on his part obviously upset Gwen Davies, for she complained in a January letter to T. J. that Maynard was inclined to take things too much into his own hands. Perhaps this was a reflection of his increasing restlessness. No one was prepared, however, for the shock which came when Maynard announced, in a letter to T. J., the joint resignation of himself and Bray, to take effect at the end of the current quarter. Reasons given were the ill health of one of Bray's two sons and the desire of both families to remove their children to an urban environment.

'You must know how much the Press has meant to me, and that I will do all I can to tide over the dislocation that may be caused . . . Do please . . . assure the Misses Davies that both Bray and myself want to do the right thing, and that the decision is not the result of any grievance.'[1]

This last statement was not entirely true, as later became apparent.

T. J. was plainly put out by the news and replied to Maynard's letter saying that he hoped they would stay on until the books already announced were completed.[2] The Davies's were on the point of leaving for a holiday in Spain, and it seems that they left Britain in ignorance of the situation. W. J. Burdon Evans interviewed Maynard who confirmed the joint resignation and also revealed that he and Bray intended setting up in partnership under their own imprint.

After much discussion between the various parties, it was finally agreed that Maynard and Bray should remain at Gregynog until the books in question were completed, which Maynard estimated would be the end of June, and that, should no suitable controller be found by that time, one or both of them would remain until September. Law was on their side because neither had a written contract of employment. As long ago as 1926, Maynard had been instrumental in obtaining legal service agreements for the senior staff but had, perhaps deliberately, omitted to do so on behalf of himself and Bray.

Dora Herbert Jones was equally stupefied at the announcement. In a letter to T. J., she revealed that she had been aware of an indefinable 'atmosphere' for some time past but was at a loss to know the cause of this sudden decision.

'To come now when everyone, inside and outside of the Press, are united to do them honour, baffles me.'[3]

Additional reasons for the resignations were not long in forthcoming when Maynard wrote a second time to T. J. His letter reveals a bitterness and rancour which had obviously been building up for some considerable time. This was directed at the owners of the Press whom he accused of avoiding him ever since the altercation with Hughes two years previously. He had been angered by Gwen Davies's sudden withdrawal of financial support towards the services of a governess which his children had shared with Dora Herbert Jones's son and daughter and also by her refusal, on seemingly trivial grounds, to allow him to use the Press car which he had planned to take on a sketching holiday in North Wales.

'I realised then . . . that, were I to remain at Gregynog, the consideration likely to be given to me would be no compensation to Mrs. Maynard and myself for living here in isolation, far away from our friends and our children[4] and cut off even from the usual amenities. Miss Davies, I believe, thinks me ungrateful, but she will never, never realise that I have made sacrifices, both personal and financial, and that she too has obligations. I hear now that I have done valuable work. Why was I not made to feel that it was appreciated? I have felt for many, many months that what I had at Gregynog was simply a "job". I asked you in my letter to assure them that I had no grievances. It is true, I haven't; the resentment has burned itself out, and I beg of you not to raise any of these questions. For one thing, it would not be fair to Bray. Neither of the Misses Davies could appreciate my point of view, and I am sick to death of battling through seas of suspicion and misunderstanding.'[5]

Misunderstanding there certainly was on both sides, and Maynard's was not the only rancour now released. Gwen Davies, writing to T. J. from Barcelona, accused Maynard and Bray of callousness and selfishness. Her pride was wounded to the point of suggesting that the Press

Board should forbid the appearance of a forthcoming article by Maynard in *The Studio*,[6] as she thought it might be full of self-advertisement. Whatever Maynard's shortcomings, this degree of vindictiveness against a man who, for so long, had been a faithful servant and an ornament to their enterprise, was surely unwarranted. One trusts that she later thought better of her outburst.

Dora Herbert Jones, ever conscious of the ambiguity of her own position, was severe in her castigation of Maynard and Bray,

'There seems to be absolutely no consideration for the Davies's . . . Their devotion to the Press comes a long way behind their devotion to themselves.'[7]

She later came to feel that her judgment had been unduly harsh.

The sudden upheaval engendered feelings of insecurity all round. Burdon Evans, and even T. J., feared that other members of the Press staff might follow suit, while Gwen Davies feared that Maynard and Bray might try to use the Gregynog imprint and suggested that it should be legally registered! Hugh Blaker's assistance was sought, and T. J. arranged a meeting in London between Blaker, Maynard, Bray and himself to discuss future staff arrangements. T. J. afterwards recorded that Blaker had made Maynard and Bray feel that they had behaved clumsily towards the Davies.

As a result of this meeting, Maynard and Bray now planned to leave the Press in the summer and, thereafter, to act in an advisory capacity, pending the appointment of their successors. They recommended that John Hugh Jones be left in charge of the Press, a suggestion which found favour neither with Gwen Davies nor T. J., as they preferred to have matters looked after by Dora Herbert Jones. Gwen Davies was adamant that a mere overseer should not be employed, as he would have to commission art work from outside. She insisted that the Press should train its own staff. She was anxious that Maynard and Bray should see *The Plays of Euripides* through the press as she foresaw that great prestige value might accrue from this publication, especially in the United States. Thereafter, she wanted no more of them and refused to entertain the idea of their services being retained in a consultants' capacity.

Adrian Boult conducted the Gregynog Choir in a performance of Bach's *St. Matthew Passion* on 14 April, for which the Press printed a programme and a copy of the libretto. Obviously work was pressing at the time, for Maynard had the text set by Linotype at the offices of *The Montgomeryshire Express*. This caused him more trouble than

AN INTRODUCTION

THERE ARE FEW WISE SAYINGS ABOUT POETRY SO FREQUENTLY ALLUDED TO & SO FREQUENTLY SERVICEABLE AS A PASSAGE FROM WORDSWORTH'S PREFACE TO THE 'LYRICAL BALLADS' OF 1805. "POETRY," HE SAYS, "IS THE SPONTANEOUS OVERFLOW OF POWERFUL FEELINGS; IT TAKES ITS ORIGIN FROM EMOTION RECOLLECTED IN TRANQUILLITY." An emotional experience, he explains, thus recollected, in this deep quietude of it was worth apparently as he complained that he had great difficulty in setting the slugs to his satisfaction.

In June the Press printed the text of a letter from Gwilym Davies, Vice-President of the Welsh League of Nations Union, concerning the *World Wireless Message of the Children of Wales*[8] for Goodwill Day, 18 May and, as a separate item, the text of the message itself. This continued to be printed at Gregynog annually thereafter.

Christina Rossetti: Poems, selected by Walter de la Mare, completed in January, was published in early April. Set in Baskerville type, it is impeccably printed in red and black on Japanese vellum. The attractive initial openings and frontispiece portrait, after drawings by Dante Gabriel Rossetti, are all by Maynard. Red hanging titles are used which could have been improved by allowing more space between them and the text. The title-page is very simple, having a swelled rule beneath the title and yet another version of the fir tree Press device.

The binding is in a Cockerell marbled paper with cream Hermitage calf spine. This leather, produced by George Russell of Hitchin, was a deliberate attempt to avoid modern injurious tanning methods and has been a favoured leather with bookbinders ever since. The firm's literature declared that,

> 'by reverting to the ancient methods, we have at last a leather as individual as calf used in the fifteenth and sixteenth centuries and, it is confidently expected, as permanent.'

It seems likely that the Gregynog Press was the first to make use of the new leather.

The special binding, designed by Maynard, is in scarlet levant morocco, decorated with a geometric design of two large ornaments bordered by five small ones, combining light and heavy gilt and blind tooling, with spine panels ruled in blind.

The book was later selected by the First Edition Club for an exhibition, held at the British Museum, entitled *Fifty Best Books of 1930*.

In May Maynard confessed that *The Plays of Euripides* would not be finished before Easter 1931. Incensed, Gwen Davies proposed not to have him back to supervise the work, as she was unwilling to 'put any more money in his pocket'. She suggested to T. J. that Bray, who had cut all the blocks for the book, should stay on to supervise the work. She pointed out that, in any case, Bray's salary was far smaller than that of Maynard! She even went so far as to suggest that Maynard

Opposite: Page from Christina Rossetti: Poems, *with portrait of Christina Rossetti, engraved by Robert Maynard.*

and Bray should be forbidden to take their holiday at the same time.

'Maynard is so abominably *astute* and so utterly lacking in any sense of obligation to his employers . . .'[9]

Poor Maynard! His seeming defection had cut deep, and it was to be a long time before he was forgiven.

In searching for suitable replacements, the Press Board cast its net far and wide. Several artists were considered or actually wrote applying for a post. They included Stephen Gooden, Donald Attwater and Douglas Percy Bliss who proposed that he might be joined, if appointed, by Edward Bawden. Gwen Davies confessed that Bliss's work made her nervous: just why was not revealed. It was, once again, Hugh Blaker who came to the rescue by suggesting the name of Blair Hughes-Stanton, the son of Sir Herbert Hughes-Stanton, R.A., a Welsh landscape painter. Hughes-Stanton was then twenty-eight years old and married to Gertrude Hermes, the sculptor. Both husband and wife already enjoyed a reputation as wood engravers. T. J. admired their work, but Gwen Davies confessed to finding it too much all of a piece and compared it unfavourably with the versatility and humorous quality of Bray's work. Hughes-Stanton's work gave her something akin to fright. She had been looking at his illustration for *Susannah and the Elders*[10] and exclaimed in a letter to T. J.,

'Oh dear! Do you think we can ever woo him from these delectable ladies? They will never do for Gregynog! That will have to be made perfectly clear.'[11]

T. J. eventually met and liked the Hughes-Stantons, as a result of which they were invited to spend the weekend of 9–12 May at Gregynog.

Joseph Thorp[12] was consulted about the list of applicants. He confided to T. J. that while he admired the Hughes-Stantons' work and that of the wife in particular, he thought that, like most artists, they would not prove competent administrators. Bliss, he considered, might be the best in that role. Bawden was supposed to be the best student to come out of the R.S.A. for twenty-five years but was also said to be 'given to lawless experiment'. Thorp counselled that, on the whole, the Bliss-Bawden partnership might be the best choice.

The weekend which the Hughes-Stantons spent at Gregynog proved mutually agreeable. The idea was mooted by Hughes-Stanton that he and his wife might be joined by William McCance,[13] a Scottish painter and sculptor, and his wife, professionally known as Agnes Miller

Parker, who was already well known as a wood engraver. T. J. later made discreet enquiries among his acquaintances concerning the McCances, as a result of which they were invited to Gregynog for a weekend later in May. Gwen Davies was obviously impressed with McCance and interested in his ideas for imparting a Celtic flavour into designs for the Press. When T. J. subsequently met McCance in London, he was quite won over, and late in May, Hughes-Stanton and McCance were duly appointed.

Hughes-Stanton was initially invited to be controller but declined on the grounds of not wishing to be involved in administrative duties, so eventually McCance became controller at a salary lower than that offered to Hughes-Stanton as artist. Hughes-Stanton undoubtedly had the greater reputation, but the reasoning behind the Board's decision was that Hughes-Stanton had two children, whereas McCance was childless! Their wives, Gertrude Hermes and Agnes Miller Parker, were both legally engaged to serve the Press as wood engravers and paid an annual retaining fee of £100 as an advance against work done. Their work was to be paid *pro rata*, and the agreed fees for blocks ranged from £2 for a block measuring 1 × 2 inches to £15 for a block measuring 9 × 6 inches.

Once again the two principal employees had no printing experience, as had happened in the case of Maynard and Bray, but the situation was a little different, for now the Press, having won acclaim in the world of fine printing, had a reputation to maintain. Dora Herbert Jones, who cared deeply about the maintenance of high standards in all that the Press undertook, was apprehensive, even to the extent of suggesting privately to T. J. that the Press Board might consider the appointment of a general foreman to oversee the printing, a suggestion which was swept aside. Instead the Board decided to send the two men away to good commercial printing firms for a short period of intensive training in the technicalities of printing and typography. McCance was thus despatched to Charles Knight and Hughes-Stanton to the Baynard Press. Their training, however, was short, lasting only about one month, and Hughes-Stanton recalls that the experience was, in any case, not of much practical benefit as virtually all his time was spent in the composing room.

The Press printed an announcement of the new appointments for distribution which is undated but which was probably printed early in June. The text is as follows:

THE GREGYNOG PRESS, NEWTOWN, MONTGOMERYSHIRE

The Misses Davies of Gregynog announce that on the retirement of Mr. R. A. Maynard and Mr. H. W. Bray from the Gregynog Press, they have secured the services of a group of four young artists, three of whom, Blair Hughes-Stanton, Gertrude Hermes and Agnes Miller Parker, are well known to collectors of wood-engravings and finely illustrated editions. The fourth member, William McCance, besides being an artist, was for three years Art Critic to *The Spectator* during the editorship of the late J. St. Loe Strachey. With the collaboration of critic and artists, the Press looks forward to maintaining its present high standard of book production and to increasing the interest already shown in the books issued from Gregynog.

The printing of *The Stealing of the Mare* translated from the Arabic by Lady Anne and rendered into verse by Wilfrid Scawen Blunt was completed on 1 July though not published until the following November. This, which is one of a cycle of tales from the celebrated medieval Islamic romance of Salame Abu Zeyd, is a splendid production in which the problems of arranging verse and prose on the same page are skilfully overcome. The text is set in Garamond type and printed on heavy Japanese paper. The title-page, which is one of the best produced by the Press, is embellished with the device printed in gold and set within a grey-blue scalloped circle. There is a magnificent frontispiece, facing the text opening, which is enhanced with gold and hand-colouring. The text opens with a decorated initial H, and there are twelve different decorated capital A's used for chapter openings, each delicately hand-coloured, four being heightened with gold. Initials and frontispiece are the work of Maynard who also designed the attractive printed paper used for the ordinary binding.[14] Here the leaf motif, which is used as a recurring theme throughout the book, appears again. The spine is Hermitage Calf. This binding is visually more appealing than the special binding which is in green unpolished levant morocco, plain apart from a gilt rule frame and spine titling and blind ties to the spine bands. The special bindings were always some two or three years behind publication. Maynard left behind no design for this cover, so that Fisher was left to his own devices when the time came to bind it.

Despite the book's considerable merits, it was rejected for inclusion in the exhibition *Fifty Best Books of 1930* organised by the First Edition Club.

There is an interesting footnote to this book. At T. J.'s request, Edward Garnett approached T. E. Lawrence with a view to getting him

Opposite: Page from The Stealing of the Mare, *printed in black, with engraved initial (hand-coloured) by Robert Maynard.* (*Reduced.*)

IN THE NAME OF GOD THE MERCIFUL, THE COMPASSIONATE!

HE WHO NARRATETH THIS TALE IS ABU OBEYD, AND HE SAITH: WHEN I TOOK NOTE AND PERCEIVED THAT THE SOULS OF MEN WERE IN PLEASURE TO HEAR GOOD STORIES, & THAT THEIR EARS WERE COMFORTED AND THAT THEY MADE GOOD CHEER IN THE LISTENING, THEN CALLED I TO MIND THE TALE OF THE AGHEYLI JABER AND HIS MARE, AND OF ALL THAT BEFELL HIM AND HIS PEOPLE. FOR THIS IS A STORY OF WONDERFUL ADVENTURE AND MARVELLOUS STRATAGEMS, & A TALE WHICH WHEN ONE HEARETH HE DESIRETH TO HAVE IT EVERMORE IN REMEMBRANCE AS A DELIGHT TASTED ONCE BY HIM & NOT FORGOTTEN.

And the telling of it is this:

The Emir Abu Zeyd the Helali Salame was sitting one morning in his tent with the Arabs of the Beni Helal and the Lords of the tribe. And lo, there appeared before them in the desert the figure of one wandering to and fro alone. And this was Ghanimeh. And the Emir Abu Zeyd said to his slave Abul Komsan, "Go forth thou, and read me the errand of this fair Lady & bring me word again." And Abul Komsan went forth as he was bidden, and presently returned to them with a smiling countenance, and he said, "O my Lord, there is the best of news for thee, for this is one that hath come a guest to thee, and she desireth something of thee, for fate hath oppressed her and troubles sore are on her head. And she hath told me all her story and the reason of her coming, and that it is from her great sorrow of mind; for she had once an husband, and his name was Dagher abul Jud, a great one of the Arabs. And to them was born a son named Amer ibn el Keram, and the boy's uncle's name was En Naaman. And when the to write an introduction, but Lawrence was not to be drawn. Garnett, in an undated letter to T. J. wrote,

'I am sorry to report that Colonel Lawrence says that £20,000 wouldn't tempt him to write a preface on any Eastern subject . . . though he might consider it for some other subject.'

The remaining books to be printed while Maynard and Bray were still at Gregynog were Charles Lamb's *Elia*, completed the previous November, and *The Last Essays of Elia*, completed 30 March 1930, the two volumes being published jointly in February 1931. Outwardly very uninspiring, within they are delightful. Set in Baskerville type and printed on Japanese vellum, they are illustrated with twenty-seven wood engravings of great charm and delicacy, executed entirely by Horace Bray from prints in the collection of Edward J. Finch. These, apart from the illustrations for *The Autobiography of Edward, Lord Herbert of Cherbury*, are unquestionably the best he ever did. The two volumes are exquisitely printed, set with ample margins and enhanced by shadow initial letters. The ordinary binding is cold and excessively dull, being covered in grey buckram. The special binding is equally uninspired, being in plain brown levant morocco with gilt spine lettering.

Announcing the forthcoming publication of these two volumes in *The Observer* for 29 November 1929, J. L. Garvin handed the Press an unexpected bouquet:

'We have remarked frequently, but not too often, that the Gregynog Press, in its line, has no superior in fastidious severity of taste. It knows a truth – "beauty when unadorned, adorned the most." Working in a sequestered nook of Wales, the Gregynog artists would be better known if they had not confined themselves to work of Welsh association. Now they have won the right to take wider ground.'

The exact date of Maynard and Bray's departure from Gregynog is not known, but it appears to have been shortly after the printing of a programme for the International Conference of the Burge Memorial Trust held there 29–30 September, together with an order of service to be used on the Sunday. This was the first of many printed orders of service for use in the Music Room at Gregynog.[15] Shortly afterwards, the two men established the Raven Press in Harrow Weald where they produced a small number of books before Bray left the Press, and Maynard continued, producing mainly greetings cards of high quality.[16]

Looking back at their work at Gregynog, it must be acknowledged

Opposite: Page from Elia, *with engraving by Horace Bray.*

The OLD & THE NEW SCHOOLMASTER

MY READING HAS BEEN LAMENTABLY desultory and immethodical. Odd, out of the way, old English plays, and treatises, have supplied me with most of my notions, and ways of feeling. In every thing that relates to *science*, I am a whole Encylopædia behind the rest of the world. I should have scarcely cut a figure among the franklins, or country gentlemen, in king John's days. I know less geography than a school-boy of six weeks' standing. To me a map of old Ortelius is as authentic as Arrowsmith. I do not know whereabout Africa merges into Asia; whether Ethiopia lie in one or other of those great divisions; nor can form the remotest conjecture of the position of New South Wales, or Van Diemen's Land. Yet do I hold a correspondence with a very dear friend in the first-named of these two Terræ Incognitæ. I have no astronomy. I do not know where to look for the Bear, or Charles's Wain; the place of any star; or the name of that theirs was a splendid achievement. They had been responsible for the production of eighteen books. One of these, *The Plays of Euripides*, was completed after they left. They had, moreover, designed and cut the illustrations, initials and decorations for all the books except one, the *Llyfr y Pregeth-wr*, which was illustrated by David Jones. In general their style was so similar that where they worked jointly on illustrations for a book, it is not easy to discern which artist was responsible for a particular engraving unless the block is actually signed. This factor contributed significantly to the unity of the books produced in this first period. There seems to be little doubt, however, that Bray was the more versatile artist of the two, a fact borne out by his illustrations for *Lord Herbert*, *The Misfortunes of Elphin* and the two *Elia* volumes. When one remembers that Maynard had received no real formal training in typography and book design, the standard he achieved is no less than amazing. Whatever the attributes of later books from the Press, the high typographical standards set by Maynard were not, in general, maintained.

NOTES

1. Letter from Maynard to T. J., dated 5 February 1930.
2. The printing of *Christina Rossetti: Poems*, though not published until 7 April was actually completed, as its colophon reveals, on 21 January and the printing of *Elia* the previous November. Thus the remaining books, already announced, were *The Stealing of the Mare*, *Last Essays of Elia* and *The Plays of Euripides*.
3. Letter from Dora Herbert Jones to T. J., dated 7 February 1930.
4. Maynard's two children had recently been sent as boarders to St. Padarn's convent school in Aberystwyth.
5. Letter from Maynard to T. J., dated 8 February 1930.
6. This was an error of fact. The forthcoming article was one by Hugh Blaker on the drawings and engravings of Maynard which appeared in *The Studio*, February 1930.
7. Letter from Dora Herbert Jones to T. J., dated 10 February 1930.
8. In later editions of the message, the word *youth* was substituted for *children*.
9. Letter from Gwen Davies to T. J., dated 3 May 1930.
10. Published in the Cresset Press edition of *The Apocrypha*, 1929.
11. Letter from Gwen Davies to T. J., dated 3 May 1930.
12. Joseph Peter Thorp, 1873–1962, journalist and print designer. Consultant to the Curwen Press *c.* 1915–1920.
13. The spelling of the name used in the text is that normally used by the artist. While at Gregynog, he used the form MacCance for typographical reasons.
14. The paper was printed by Hodgson, using a combination of three small blocks.
15. The Music Room, equipped with an organ and grand piano and used for concerts and services, was never consecrated as a chapel.
16. For further information about the press the reader is referred to the late James A. Dearden's article entitled 'The Raven Press' in *The Private Library*, Second Series, volume 6:4, Winter, 1973.

10

Hughes-Stanton and McCance

HUGHES-STANTON, McCANCE AND THEIR FAMILIES were, by October 1930, living in Tregynon, in the houses which were formerly occupied by Maynard and Bray, and the two men were working in the Press, but the atmosphere was far from settled. All was not well with Hughes-Stanton's marriage in which a rift had already appeared before they came. His wife was not happy in the country, far removed from ready sale outlets for her sculpture. Both Hughes-Stanton and McCance were irked by the necessity of completing *The Plays of Euripides* and anxious to embark on work of their own design. Dora Herbert Jones became, alas, all too often the butt of their displeasure, partly because she was daily in close proximity and partly, perhaps, because they sensed her reservations concerning their work. A letter from Gwen Davies to T. J., written on 22 October, gives some indication of the state of things:

'Idris wants more wages, John [Hugh Jones] a career.[1] Dora has threatened to resign. Stanton has a violent temper . . . *Fisher* is a darling . . . Keeps things level in the bindery. There has been a lovely "flare-up" because W.J.'s [Burdon Evans] "J" has been left out of the list of directors on the new notepaper!!! Put that in your pipe and smoke it. What a lovely sketch it would all make . . . One gets alarmed at first, then cross at the pettiness of it all – then tremendously tickled and amused at the absurdity!!! Of course, poor old Hughes is suspected of being at the back of it all!!!!'

A contretemps arose over McCance's first piece of printing. This was an order of service for a conference of the Advisory Education Committee of the Welsh National Council of the League of Nations Union which was to be held at Gregynog 14–17 November. Dora Herbert Jones, who assisted Gwen Davies with the preparation of all the services, was accused by McCance of having made the service too long. Early in November, he declared that the text was too extensive for him to get printed in time and that Mrs Herbert Jones should have edited and cut it down to size 'as if', Gwen Davies expostulated to T. J., 'it were a newspaper'.[2] In desperation, Gwen Davies posted the text to

T. J. with a plea for him to get it printed in London. This instruction was immediately countermanded by telegram; McCance had reconsidered. Finally the service, together with a list of the conference delegates, was printed under McCance's direction. Both items bear the colophon,

'Planned by William MacCance. Engraved by Blair Hughes-Stanton, and carried out by the staff of the Gregynog Press.'

The lower covers of each display a new Press device, a large snail shell, designed and cut by Stanton above the tag 'Levavi Oculos'[3] and then the name of the Press in its oft disputed form 'Gwasg-y-Gregynnog'. This was not allowed to pass without comment from W. J. Gruffydd who was against the form and Ifor Wiliams[4] who was in favour!

Relations continued strained between the controller and artist on one hand and the secretary on the other. While Dora Herbert Jones greatly admired Hughes-Stanton's artistry, skill and vision, she was apprehensive concerning McCance's ability to handle typographical matters. On the other side lay a watchfulness lest her unease, clearly apprehended, should communicate itself to the Davies's. Neither side fully trusted the other, and both were nervous of their position. This was certainly so in the case of Dora Herbert Jones who was extremely sensitive, and the sisters did little to reassure her, though it is obvious from Gwen Davies's letters to T. J. that she valued the secretary and her services very highly indeed.

That the secretary was not alone in feeling uncertain of the typographical capabilities of the two new recruits was made plain in a letter which Gwen Davies wrote to T. J. on 22 November in which she referred to this very matter. She suggested that they should be sent away to 'Practise and learn under expert guidance'. The names of Emery Walker and Bruce Rogers were mooted as possible tutors, but nothing came of her plan. Her unease is even more apparent in a draft letter which she addressed to McCance after seeing the proof of a Christmas card. Implying that their work was not up to the standard required, she even went as far as to point out that typographers of quality had been overlooked in favour of their appointment. Mercifully, T. J., with characteristic tact and diplomacy, persuaded her of the wisdom of withholding the letter which might well have caused an early rupture in relations between the owners of the Press and their principal employees.

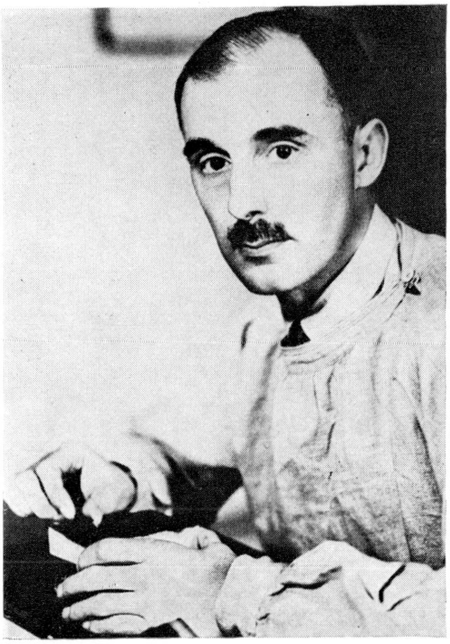

Robert Maynard, late 1920s.

Horace Bray, 1923.

George Fisher, 1928.

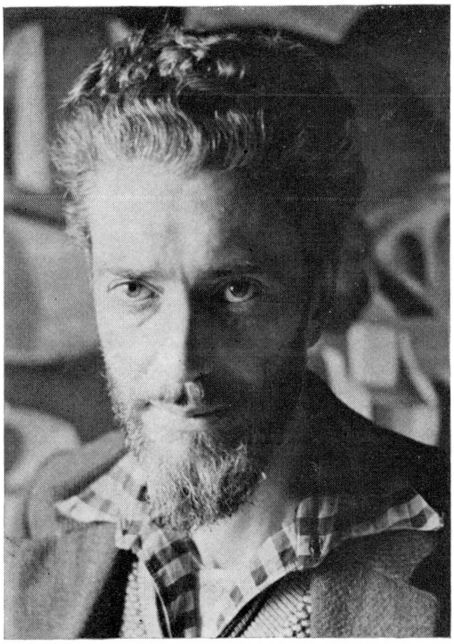

William McCance, 1930s.

Gertrude Hermes, c. *1933*.

Blair Hughes-Stanton, c. *1932*.

Agnes Miller Parker, c. *1930*.

The fact that two books only had appeared from the Press this year rankled with Gwen Davies and added, unfairly, to her irritation with the two men. When, in late October, McCance had sought permission to purchase some small matrices from the Monotype Corporation, at a cost of £250, his request had not been granted. The incident provides a good example of the Davies's parsimonious attitude on occasion. While they always stated that it was unnecessary for the Press to make a profit, nevertheless, they were unhappy at the prospect of its making a loss, as seemed more than likely this year.

There are three variant Christmas cards for 1930. Two bear, on the front, a wood engraving of a dove carrying an olive branch, a new version of the fir tree device and the imprint 'Gwasg.Y.Gregynnog'. The cards are identical except for the greeting which also appears on the front. One is addressed from 'the Misses Davies' and the other from 'the Gregynog Press'. Inside are three wood engravings by Hughes-Stanton and some lines from Robert Bridges's *The Testament of Beauty* underlined with a double rule. The total effect is somewhat cluttered. The third card is smaller and bears within a simple message of greeting from Mrs Edward Davies and the Misses Davies, embellished with an engraved initial letter and a sprig of holly at the foot. All three are printed on Japanese vellum. On each the greeting is underlined with a double rule which strikes one as an unnecessary piece of artifice.[5]

Life at the Hall had been enlivened for the Davies's during the autumn in entertaining several interesting guests. These included George Bernard and Charlotte Shaw, who came to lunch in September, and Miss Elizabeth Haldane, Frau Toni Senders[6] and Albrecht Mendelssohn-Bartholdy, Mendelssohn's grandson, all of whom attended the Burge Memorial Trust International Conference held in September.

Among the guests at the League of Nations Union conference in November was Salvador de Madariaga[7] of whom Dora Herbert Jones reported,

> 'Madariaga was delightful. [He was] the spark among the stubble. His mind was like quicksilver.'[8]

At this time, T. J. was trying to negotiate publishing rights for an edition of Madariaga's *Don Quixote* which was already contracted for by the Oxford University Press. On account of the prior arrangement, Madariaga tried to persuade the Press Board to publish his *Sir Bob* in preference. Having once been convinced of the desirability of a course of action, T. J. was not one to be easily turned aside from it, and it says much for his powers of persuasion that he actually managed to secure a somewhat reluctant agreement from O.U.P. to enable the Gregynog Press to publish *Don Quixote* first in a limited edition.

In other respects the Davies's had had an unsettling year, events culminating in a sad estrangement from their brother who, in the autumn, had invited them to take over a large part of financial responsibility for running the family estate at Plâs Dinam. This they turned down on the grounds that his offer came ten years too late: now they had too many commitments at Gregynog. Acrimonious correspondence followed which wounded the sisters deeply. In November, David Davies took lease of a large house near Colchester and moved out of Plâs Dinam. In a desperate attempt, as he saw it, to save the family fortunes, he decided to sell off some of the estate farms, a move likely to cause hardship to the tenants which further upset Gwen and Daisy. Indeed, so troubled were they by this last move that, the following year, they were to advance money to some of the worst hit tenants.

'We are having to help some of the poor people at home by advancing money on their farms – it is very pathetic – servant girls withdrawing their hard won savings.'[9]

Here was the other side of the coin. Their basic generosity of spirit is revealed in this act of feudal loyalty to old tenants and retainers.

Early in 1931, plans were afoot to publish Robert Vansittart's *The Singing Caravan* with illustrations by Gertrude Hermes. When Hughes-Stanton announced that, as his wife did not wish to participate, he proposed to do the engravings himself, Gwen Davies had to accept, but clearly she was still not sure of the artist, for, in reporting the incident to T. J., she wrote,

'He can't go far wrong with camels anyway!'[10]

For some time the Press had been negotiating with W. H. Davies to publish a collection of his poems. The book, eventually published as *The Lovers' Song-Book*, was to have a chequered career. The agreement had been that Davies should provide thirty poems for a fee of one hundred pounds, plus a royalty of five per cent on sales, on condition that the poems would not appear elsewhere for twelve months after publication. In December 1930, the secretary had complained that the poet had given short measure in that he had supplied only twenty-eight poems and remarked with characteristic acerbity, 'and they are thin'.[11] She also pointed out that two of the poems had already been published. Clearly she did not feel that Davies was playing fair and recommended that he be requested to supply four further poems!

Hughes-Stanton and McCance had been hard at work on the book for the past few weeks, but the initial response to the production was lacking in enthusiasm, to say the least. Dora Herbert Jones took exception to the typography, and Hugh Blaker, when consulted by Gwen Davies, admitted that he did not like the engravings. Eventually, McCance was instructed to invite Stanley Morison to come to Gregynog for a weekend in February to give his opinion. Morison, however, declined to come and further upset Gwen Davies by the manner of his reply to McCance which, after apologising for being unable to come, went on,

'much as I should like to come down and perturbate the Calvinists and put my foot on the neck of a few Welsh heretics and perhaps an occasional heresy of your own!'[12]

Relations between Morison and the Press obviously became strained, for a Press Board Minute of 23 October 1931 recorded,

'Resolved that no further volumes of the Gregynog Press be sent to Stanley Morison . . . unless he definitely orders them, and that no statement of account be sent to him for those already in his possession.'

Presumably the Board considered his services as informal consultant either insufficient or, alternatively, his pronouncements too damning to warrant the continued supply of free copies.

When T. J. was shown a proof copy of *The Lovers' Song-Book* a week or so later, he condemned it out of hand, and further work on it was postponed, greatly to the chagrin of Hughes-Stanton and McCance. Looking at the book now, after a passage of more than forty years, one is inclined to think that the Press Board made a sad error of judgment. It is infinitely more interesting than the unillustrated version which finally appeared nearly three years later. Set in Poliphilus type and printed on Japanese vellum, with sepia-coloured wood-engraved initials and tail-pieces by Hughes-Stanton and double rules down the gutter and along the tail, it makes an interesting contrast with previous Gregynog books. Certainly it has faults: there is a marked lack of harmony throughout, some pages being much more successful than others. Some of the longer lines are eccentrically divided, the second half being ranged on the right of the line below, the remainder of the line being filled up with a lower-case Garamond swash e. Moreover, some of the vertical tail-pieces do not sit happily on the page. Hughes-Stanton's historiated initials and illustrations are, however, original and attractive, and one cannot help feeling that it would have been preferable to have worked at amending the book than to jettison it altogether.

Though the Board rejected the book ostensibly on the grounds of inferior typographical layout, there is evidence that the members found certain of the illustrations too erotic!

In March the Press sent out an invitation to subscribers and others to attend an exhibition of books from Gregynog and other presses which opened on 9 March at John & Edward Bumpus Ltd., 350 Oxford Street, London. There is no record extant of exactly what books were shown.

Maynard returned briefly to Gregynog in February and again in June to supervise the printing of *The Plays of Euripides*, translated by Gilbert Murray. The printing of the second volume was completed on 27 June, and the work was published three weeks later. Alas, the delay in the production of this book, so long awaited by the trade, adversely affected its sales. Five hundred copies[13] were printed, and it was by far the most expensive Gregynog book ever produced. By 1931 the fine book trade was in the doldrums due to the slump, and the delay provided booksellers with a heaven-sent opportunity to reduce their orders. Bumpus, for example, reduced their order from sixty copies to twenty, and Random House, their main American distributor, reduced theirs likewise from one hundred and twenty-five to fifty copies.[14] The gravity of the situation caused the Press Board to reduce the sizes of forthcoming editions, and most of the later editions did not exceed two hundred and fifty copies.

Many copies of *The Plays of Euripides* were later given away to colleges and libraries in Wales, and even after the Press closed, there were copies left. No less than sixteen of the twenty-five special levant copies were finished as late as 1952 by John Ewart Bowen at the National Library of Wales. It was a sad fate for a book which an *Observer* review described as 'a magnificent folio edition' and 'a fine example of modern book making'.[15]

Despite that, the book proved a fitting swan song for Maynard and Bray. Set in Bembo and Fairbank Italic, with wide margins, and printed on a special make of Batchelor's hand-made paper, the two volumes are

Engraving by Horace Bray from The Plays of Euripides.

extremely handsome. The numerous wood engravings were drawn by Maynard and Bray, from fifth century Greek vase paintings, and cut by Bray. The only colour in the book is the red in which the title-page device is printed.

The brick-red cloth used for the ordinary binding is of interest in that it was a very early use of James Morton's Sundour non-fade book cloth, developed only shortly before for the binding of the collected works of George Bernard Shaw. In order to save time, Fisher bound the two volumes as one for the special edition which is covered in brown levant morocco, blocked in gilt on each cover with Greek figures taken from the engravings. On the lower cover, the figure is enclosed in bands of vertical gilt lines.[16]

NOTES

1. John Hugh Jones was currently considering entering journalism.
2. The incident was reported later by Gwen Davies to T. J. in a letter dated 22 November 1930.
3. Also used on earlier concert and conference programmes.
4. Ifor Williams, lecturer in Welsh, University College Bangor.
5. Gwen Davies was later to claim that these cards had lost the Press a lot of subscribers, though there is no evidence to prove that this was so.
6. The first woman member of the Reichstag.
7. Salvador de Madariaga, 1886-1978, Professor of Spanish Studies at Oxford, late Director of Disarmament, League of Nations Union Secretariat, and author.
8. Letter from Dora Herbert Jones to T. J., dated 17 November 1930.
9. Letter from Gwen Davies to T. J., dated 10 June 1931.
10. Letter from Gwen Davies to T. J., dated 14 January 1931.
11. Letter from Dora Herbert Jones to T. J., dated 12 December 1930.
12. Letter from Stanley Morison to McCance, as quoted in a letter from Dora Herbert Jones to T. J., dated 2 February 1931.
13. Only *Poems by Henry Vaughan* and *Detholiad o Ganiadau gan T. Gwynn Jones* were issued in such large editions.
14. Later the same year, Random House cancelled *all* orders for Gregynog publications.
15. Reviewed by J. L. Garvin in *The Observer* on 9 December 1931.
16. The design of the binding on the copies later bound by Bowen differs slightly.

11

Four Engravers

THE ONLY OTHER BOOK PUBLISHED DURING 1931 was Milton's *Comus*, completed in early June and published in October. Due to the fact that Hodgson was still engaged on printing *The Plays of Euripides* on the platen press, *Comus* was printed by Idris Jones on the Albion hand-press. The colophon states that Hughes-Stanton collaborated in the design of the book and its printing. The narrow format suits admirably Hughes-Stanton's lovely costume engravings which are a marvel of fine detail and superbly printed on Japanese vellum. Unfortunately, it suits the text setting, which is in 12-point Baskerville, less well; the regimented columns of type become oppressive.

The ordinary binding has a wide yellow buckram spine piece, lettered on spine and cover, and buff coloured paper sides. Hughes-Stanton had no experience of bookbinding design, but, under Fisher's tutelage, he rapidly learned to adapt his designs to the requirements of a book cover.[1] His first design for the special binding of *Comus* is restrained but attractive. The dun coloured unpolished levant morocco cover is decorated with broad perpendicular panels of fine blind lines, contrasted with broader gilt lines which are carried over to the turn-ins and broken on the front cover by a blind title panel.

The special bindings, hitherto designed by Maynard or Bray, were simple, almost austere bindings which relied for effect on the beauty of the colour and quality of the leather which were displayed to greater advantage by the lack of decoration. These had required of Fisher nothing that any skilled finisher would not have found well within his capabilities. With the advent of Hughes-Stanton and McCance the concept of binding design changed drastically. Gilt ruling on board edges and gilt dots on head- and tail-caps now disappeared, but raised spine bands normally remained. Fisher greatly admired Hughes-Stanton as an artist and wood engraver but was less enthusiastic concerning his skill as a designer of bindings:

'They interested me more than they pleased me in that, while they afforded me ample opportunity to display my skill as a finisher, I considered them more suited to the decoration of the printed page than to a tooled leather cover.'[2] Despite Fisher's reservations, the six Hughes-Stanton bindings remain for many collectors the most exciting and sought after of all the Gregynog bindings.[3] Two of the special bindings were designed by McCance and these, as we shall see, are of particular interest. All are signed with the names of both designer and binder.

Ephemeral pieces of printing emerged at regular intervals, too regular, it would seem, for Hughes-Stanton and McCance who regarded their preparation as an unwelcome interruption to more interesting work. The earliest of these was a handsomely printed twenty-eight page subscription list for the Elphin Lloyd Jones Memorial Fund, bound in red buckram. Among other items which followed were three prospectuses, four orders of service, a programme for an exhibition of work produced in the distressed areas of South Wales which opened at Gregynog on 12 August, two concert programmes and the libretto for Ralph Vaughan Williams's *Mass in G Minor for Soli and Double Chorus*, produced for use by the Gregynog Choir who performed the work on 11 July. On these pieces two more press devices appeared, both of them designed by Blair Hughes-Stanton. Small wonder that the minutes of a Press Board meeting held in September recorded a decision to ask John Hugh Jones to look after visitors to the Press in order that the work of controller and artist might be undisturbed! Yet life was not all nose to the grindstone, for Hughes-Stanton at least who, to the same meeting, reported on a visit he had recently made to Paris where the First Edition Club were staging an exhibition which included items from the Gregynog Press. These, Hughes-Stanton had been disappointed to find, had been poorly displayed.

During the autumn of 1931, an appeal was launched for funds to erect a village hall in Ponterwyd, a village nestling at the foot of Plynlimon in Cardiganshire, in memory of a native son, Sir John Rhys.[4] Gwen Davies generously agreed to foot the bill for printing the appeal at Gregynog. This twelve-page booklet, set in Poliphilus type, includes two engravings by William McCance, a frontispiece portrait of Sir John and a view of his birthplace, Abercaero. McCance, inspired by the nature of the enterprise, offered to sell signed copies of the portrait to swell the funds and to waive his right to twelve free prints. It was a generous gesture which, alas, did not meet with a great deal of success.

Work was well advanced on W. J. Gruffydd's *Caniadau*, which the

Opposite: Title-page to Comus, *with engraving by Blair Hughes-Stanton.* (*Slightly reduced.*)

COMUS

A MASK BY JOHN MILTON

WITH
A FRONTISPIECE AND
THE SIX CHARACTERS IN
COSTUME DESIGNED AND
ENGRAVED ON WOOD BY
BLAIR HUGHES - STANTON

THE GREGYNOG PRESS
MCMXXXI

Press had hoped to publish during 1931, when problems arose due to the decision to use Eric Gill's Perpetua type which had recently been put on the market by the Monotype Corporation. The book as designed required the use of italic, and by December, the italic version of Perpetua was still not available. Finally, it was decided, with the author's approval, to go ahead with the printing, using roman, printed in red, where the italic was to have been, but publication was thus delayed until July of the following year.

Agnes Miller Parker, who had recently spent a great deal of time working on wood engravings for *The Fables of Esope*, grew dissatisfied with the payment for blocks initially approved by the Press Board. In October an interview took place between the artist and W. J. Burdon Evans. The latter was sympathetic to her plight when he discovered that, up to September, she had spent more than eight months on continuous work. During this time she had produced twenty-nine blocks for which she was entitled to payment of £167. The problem stemmed from the fact that Hughes-Stanton had negotiated the original charges for blocks based on the timing of his own work, and he was an unusually rapid engraver. The opinions of several publishers who regularly employed wood engravers were sought by T. J. The replies are revealing in that they indicate the current valuation placed on the work of such artists by the book trade in general. Ivor Nicholson of Nicholson and Watson declared that the agreed terms were already too high and begged T. J. to defer any re-negotiation. A representative of William Heinemann claimed that the prevailing fee was adequate as the market for wood engravings was slow. He opined that the award of an additional ten to twenty per cent would be generous, pointing out that publication by the Press served as a valuable advertisement for the artist! The Press Board obviously felt the need to act none too precipitately and with circumspection, as any increase paid to Agnes Miller Parker would have to apply equally to Gertrude Hermes who had signed a similar agreement.

Gwen Davies was in favour of awarding a twenty per cent increase on payment for blocks, but, after all his initial sympathy, her financial watchdog was prepared to offer only ten per cent. The original agreement with the two women artists had been for the term of one year which expired at the end of September. In December the Press Board decided not to renew the agreement but to employ them on a freelance basis in future. Agnes Miller Parker wished in any case, Burdon Evans

Opposite: Title-page to The Fables of Esope, *with engraving by Agnes Miller Parker.* (*Reduced.*)

THE FABLES OF ESOPE

TRANSLATED OUT OF FRENSSHE IN TO ENGLYSSHE BY: WILLIAM CAXTON

WITH ENGRAVINGS ON WOOD BY AGNES MILLER PARKER

THE GREGYNOG PRESS: NEWTOWN
MONTGOMERYSHIRE: - MCMXXXI

claimed, to give up wood engraving and concentrate on painting, but one suspects that she may well have been wearied by the Board's parsimony. Whatever her feelings, she agreed to the new arrangement.

The decision to produce a straight, unedited reprint of Caxton's *The Fables of Esope* led to an altercation between controller and chairman. The printing was completed on the last day of November, and when the proofs were shown to members of the Board on 14 December, T. J. took exception to the annotation which McCance had seen fit to add to the colophon. The offending note read,

'The Printer felt that the original savour of the Fables could be retained more truly by following the early manner of spelling. There has been no attempt to re-edit the original text and only a word here and there has been altered where an obvious misprint made the meaning obscure. The original / used by Caxton was found to be disturbing with modern [*sic*] Bembo type and was replaced by the colon.'

T. J. felt that what he considered to be ephemeral comment had no place in a printed book and wrote to McCance,

'There are a lot of things you must have "felt" about the printing of the book, but when it goes out you let it speak for itself and accept the public verdict . . .'[5]

He went on to demand the removal of the note which, of course, would necessitate the reprinting of the whole sheet. An obdurate reply was sent by return in which McCance referred to his having sought the opinion of 'one who shall remain anonymous but whose name is, more than any other, a household word in book production, typography and publishing'. (An obvious reference to Stanley Morison with whom McCance corresponded frequently at this time.)

'. . . he said that, in the circumstances, it was not just inoffensive but *absolutely essential*. So if the Directors wish the colophon reprinted, after my explanation, then the order to Hodgson must come from someone other than myself, for I should be acting against my own conscience and judgment were I to delete the note.'

The letter, of which McCance carefully sent a copy to each of the directors, went on to say that, in his opinion, the text should have been re-edited and that, as the Board had decided against so doing, he believed that it was only fair to prospective purchasers to make the point clear. The Press, McCance declared, ought to have a literary adviser and ought not to rely on the controller to determine such matters. Impasse appeared to have been reached. In the end both sides had to retract somewhat, and the colophon note was reprinted as follows: 'In this edition obvious errors in the original text have been corrected. The diagonal / used by Caxton was found to be disturbing with the type used and was replaced by the colon.'

For Christmas 1931 three different greetings cards were printed. The smallest is printed on a single deckle-edged grey card and bears merely the greetings message from Mrs Edward Davies and the Misses Davies. The other two are larger and have identical wrappers, except for colour, printed with a greeting from the sisters, in the centre of which is one of the two recently designed Press devices. Each version encloses a single sheet, the blue wrapper fourteen lines from Spenser's *Hymn to Eternall Love*, printed on hand-made paper, and the buff wrapper twenty-two lines from Drinkwater's *The Fires of God*, printed on Japanese vellum, in which the two verses open with large wood engraved initials designed by William McCance. These are placed in close juxtaposition, one above the other, so as to form a kind of symbolic cypher.

Some of the previous year's fears concerning the Press had been dispelled by the publication, during the year, of three books – the *Elia* volumes, *The Plays of Euripides* and *Comus*. All of these had been well received, and more books were already in active preparation. Meanwhile, Gregynog Hall was gaining a growing reputation as a conference centre of some importance in Wales.[6]

On a personal level the year had not been without its problems for the Davies's. To their chagrin, they were still at variance with their brother over matters concerning the family estate. In the spring and again in the autumn, Daisy Davies had spent several weeks at a clinic in Grasse, undergoing ultra-violet ray treatment for a severe throat condition which was to be a continuing disability for some years to come. Interesting visitors continued to drop in at Gregynog from time to time. Adrian Boult came again following his appearance in May at the Montgomeryshire County Music Festival at Newtown, bringing with him Isobel Baillie and Keith Falkner. Boult was also present at the July conference of the National Council of Music at which Gustav Holst made his first appearance at Gregynog. During the weekend, Holst conducted the choir in madrigals and a group of folksongs, arranged by himself and his friend, Ralph Vaughan Williams, at a concert of chamber and vocal music held on 11 July. Though Eric Gill never paid his promised visit to Gregynog, in July the Press was visited by his daughter, Joanna, and her husband, René Hague, accompanied by Gill's secretary and biographer, Donald Attwater who, it will be remembered, had applied for the post of controller of the Press a year earlier. Naomi Mitchison and her husband, who were friends of Gertrude Hermes, also paid a visit to the Press in September.

By the time 1932 dawned, McCance and Hughes-Stanton were well into their stride, but a general atmosphere of unease still prevailed in the Press. In January it was discovered that, in printing the prospectus for *The Fables of Esope*, several characters of the wrong fount had been inadvertently used, with the result that the Board ordered it to be reprinted. The wood engraved blocks used were replaced by two more, so as to guard against undue wear and tear in the reprinting. Due to the fact that the compositors were suffering from influenza the type for an order of service to be used during the Conference on Social Reconstruction in the Distressed Areas of South Wales and Monmouthshire, held 15–18 January, was set by Western Typesetting Services of Bristol. It seems likely that the printing staff also became afflicted, as the programme for the conference was printed in Newtown by the Montgomeryshire Express Ltd.

Shortly afterwards, a situation arose concerning the printing of W. J. Gruffydd's *Caniadau*. The copy had been set from an earlier edition and the proofs sent to the author who was dilatory in returning them. With unwise lack of caution the Press had gone ahead, without the author's approval, and printed fourteen of the fifteen formes. When the corrected proofs finally arrived in late February, a great many corrections had been marked for which, in a somewhat tactless letter, McCance blamed the author. Gruffydd, in self justification, replied that, in setting from a previous edition, McCance, not being Welsh speaking, could not have realised that the bulk of the alterations derived from the new system of Welsh orthography introduced within recent years. The reprinting of the book took so much paper that it was feared not enough would be left for the printing of a prospectus. Later in the summer, a prospectus did appear, but it was restricted to a single sheet.

Since Maynard's departure, John Hugh Jones had been unsettled. Doubtless he felt that, after nine years at the Press, he ought to have achieved a position of greater authority and responsibility. Some months earlier, T. J. had dissuaded him from entering upon a career in journalism as he felt that Jones's talents did not truly lie in this field. He had also tried to join the staff of the BBC in Wales but had been

Opposite: Page from The Fables of Esope, *with engraving by Agnes Miller Parker.* (*Reduced.*)

A thousand wyles haue I sayd the foxe: For I haue a sak ful of scyences and wyles: And I am so grete a clerke: that none maye begyle ne deceyue me: And as they were thus spekyng to gyder the cat perceyued a knyght comynge toward them: whiche had many dogges with hym: and sayd to the foxe: My godsep:

certaynly I see a knyght comynge hyther ward: whiche ledeth with hym many dogges: the whiche as ye wel knowe ben our enemyes: The foxe thenne ansuerd to the cat: My godsep: thou spekest lyke a coward: and as he that is aferd: lete them come and care not thow: And Incontynently as the dogges perceyued and sawe the foxe and the catte: they beganne to renne vpon them: And whanne the foxe sawe them come: he sayd to the kat: Flee we my broder: flee we: To whome the kat ansuerd· Certaynly godsep: therof is none nede: neuer the les the foxe bylued not the cat: but fledde: and ranne as fast as he myght for to saue hym: And the catte lepte vpon a tree and saued hym self: sayenge: Now shalle we see: who shalle playe best for to preserue and saue hym self: And whanne the catte was vpon a tree: he loked aboute hym: and sawe how the dogges prevented by lack of experience. T. J. now offered to enquire of Walter Lewis at the Cambridge University Press and John Johnson at the Oxford University Press whether there might be any opening for him in either establishment. Neither was able to hold out any hope of employment, and Jones reluctantly resigned himself to remaining at Gregynog.

For some time the Press had been working on an edition of Gilbert White's *Natural History of Selborne* for which Gertrude Hermes had been commissioned to provide illustrations. Learning that, so far, only the frontispiece and two other wood engravings had been completed, Burdon Evans began to suspect that the artist was deliberately prolonging the task. He was not to know that her creative ability, at that time, was virtually anaesthetised by personal misery, a fact revealed in a March letter to Gwen Davies announcing that she had left her husband and had taken a flat in London where she proposed continuing her work for the Press.[7] Burdon Evans pressed the termination of her contract which was, in any case, due for renewal in June.

The Fables of Esope was published on 2 April. Set in Bembo type and printed on a paper specially made for the Press by Barcham Green, the text is illustrated with thirty-seven large wood engravings by Agnes Miller Parker. Never can living creatures have been more successfully portrayed in so static a medium. The feeling of movement is so admirably achieved that the engravings have an almost three-dimensional quality. Sadly, their effect is marred in that they are too closely crowded by type. McCance's historiated initials are over large, and, where they appear close to the illustrations, serve only to fragment the page. There is a beautiful and apposite wood engraved title-page. A few of Agnes Miller Parker's engravings turned out to be too heavy for the type, and Hodgson lightened the ink by mixing with white. The presswork, on the roughish hand-made paper, appears faultless and was much admired. Walter Lewis, writing to T. J. on 17 May, comments,

> 'I do not think I have ever seen such a beautifully printed book . . . an example of the finest presswork I have seen. I only hope that I shall one day be able to do a book equal to it.'

For the first time sheepskin, obtained locally from Lewis's tannery in Newtown, was used for the binding. Here it is used in its natural colour, and the spine is lettered in black. McCance was responsible for the design of the special binding, executed in biscuit coloured polished

Loyd Haberly, 1939.

James Wardrop, 1930s.

Herbert Hodgson (second from left) and The Venetian Dance Band, 1930s.

Gregynog bindery interior, late 1920s.

levant morocco. The front cover is embellished with an ingeniously contrived interlacing of large letters to form the word ESOPE which are onlaid in reddish brown niger, outlined in gilt. The remainder of the title is tooled in blind, and the author's name is in outline gilt letters, the whole enclosed within concentric ruled panels. The outer ruled frame continues to the lower cover where it terminates in a small rectangle which encloses the Press device. The spine title panel displays the title in gilt, the author's name in blind together with eight square dots, onlaid in red and outlined in gilt, within a double rule gilt frame. At the foot of the spine, the publication date is shown, tooled in gilt roman numerals. The total effect is one of complete harmony. It seems highly likely that had McCance further pursued the field of bookbinding design it might have proved to be his *forte*.

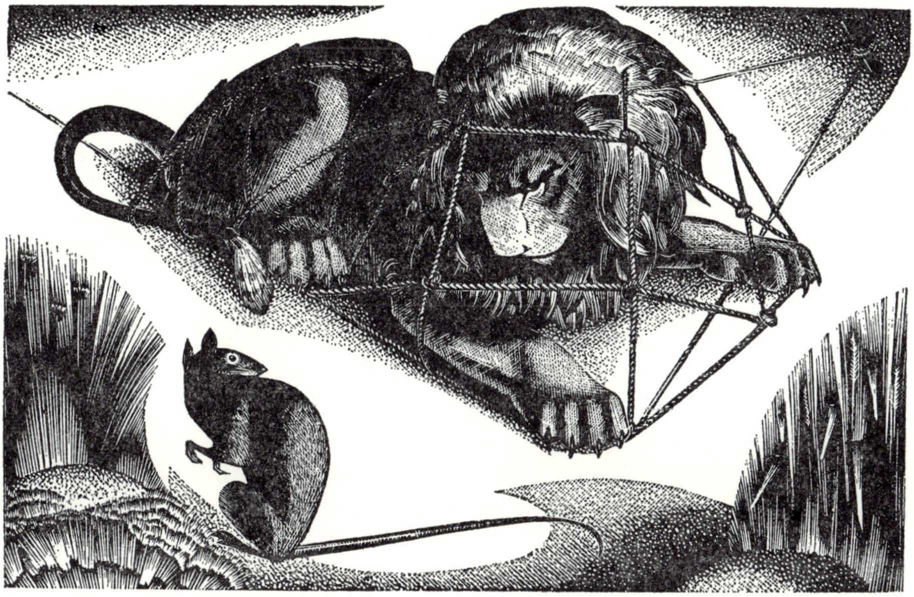

Engraving by Agnes Miller Parker from The Fables of Esope.
(*Slightly reduced.*)

For a concert of vocal and instrumental music held on 11 April the Press printed a programme and a separate item recording the dedication to Adrian Boult of a performance by the Gregynog Choir of Ralph Vaughan Williams's *Benedicite*. The composer made his only appearance at Gregynog to conduct the work in which Elsie Suddaby sang the solo part, in addition to songs by Purcell, Boyce, John Attey and John Bartlet. Dora Herbert Jones was largely responsible for the visit as she and the composer had been acquainted for some years in her capacity as a folk singer. Indeed, he had arranged several settings of folk songs specially for her. Vaughan Williams, however, did not find Gregynog to his taste and future invitations were politely refused.[8] The concert opened with Sir Walford Davies playing three choral preludes for organ by Brahms, Herzogenberg and Bach. He was later joined by Evelyn Cooke in Beethoven's *Sonata for Piano and Violin in A Major* and Vaughan Williams's *The Lark Ascending*. The remainder of the programme was made up with songs by Peter Warlock and Walford Davies and folk songs rendered by the choir.

In May Blair Hughes-Stanton came under attack when the Press Board took exception to the manner of his domestic life. It was suggested that his contract be terminated in the coming September, but that his services should be retained to oversee the completion of *The Revelation of Saint John* on which work had already begun. Though to-day it would seem extreme to dispense with the services of so talented an artist on such an irrelevant pretext, one is compelled to remember that public opinion was very much narrower in outlook in the 1930's and that the village community of Tregynon, where the majority of the Press staff lived, was then, as now, a small and tightly knit one where the lives of the villagers were closely interwoven. Nothing could be allowed which might taint the reputation of the Press. Hughes-Stanton, not unnaturally, resisted a termination of his contract without adequate financial recompense and managed to hold on to his appointment until the following year.

One small item of interest was printed during the summer. This was a handsomely designed reprinting of two specimen pages from Goethe's *The Sorrows of Young Werther*, from the translation by William Rose originally published by the Scholartis Press, for a Goethe exhibition held in Leipzig. It is set in Perpetua type and printed in black and tyrian red on Japanese vellum. Also shown at the exhibition was a volume of Goethe's works, specially bound by George Fisher to a design by William McCance.

In June Brooke Crutchley, recently appointed as typographical assistant to the Cambridge University Press, paid a visit to Gregynog. He afterwards wrote to T. J. and expressed admiration of what appeared to him a successful commingling of the aesthetic with sound economic sense. (Not a view that everyone would have shared!) In many ways the Press Board were concerned with the economic aspects of the venture. At a meeting of the Board, held in July, they refused to sanction the expenditure of two hundred pounds quoted by the Monotype Corporation for the cutting of a new type face designed by McCance. At the same meeting the controller complained about the constant alterations made by the Board to the scheduled work programme which he claimed interfered with his creative work. This referred to the steady stream of ephemeral printing which the Press was called upon to produce, often at very short notice. The Chairman's response to this was to suggest that McCance might care to relinquish his post!

Relations between management and staff were always somewhat strained, due largely to the fact that the directors were too remote from the day to day working of the Press. As a result, there were frequent misunderstandings on both sides, and all too often situations involving personnel were mishandled. The next to come under fire was Agnes Miller Parker whom the Press accused of discourtesy. This arose from her acceptance of a commission to illustrate a book for the Golden Cockerel Press. As her contract had been terminated, she was somewhat taken aback by their outburst and wrote to T. J. reminding him that while she had promised to give preference to commissions from the Board, of late they had not seen fit to call upon her services. She added that if the directors felt some 'parental' interest in her other commissions she would keep them informed but that she hardly felt this was likely in view of the way in which they had treated her earlier request for increased payment. She went on,

> 'This attribution to me of discourteousness comes at a very inopportune moment when I feel very deeply the unsympathetic treatment on the part of the directors to my friend, Mrs. Hughes-Stanton, whose suffering could, with this latest knock,[9] have led her to a complete bitterness with the world, had she not been the fine creature that she is. I suppose a company has to become benumbed to the finer humanistic values.'[10]

T. J.'s reaction to this is not preserved.

Adrian Boult presided over a second concert on 16 July. The choir sang motets by Hubert Parry and Bach, Elsie Suddaby sang pieces by Schubert, Purcell and Bach. Sir Walford Davies played the adagio and rondo from Beethoven's *Sonata for Piano in B Flat. Opus 22,* in addition to organ pieces by Brahms, Mendelssohn and Handel, and the finale was another rendering by the choir and Elsie Suddaby of Vaughan

Williams's *Benedicite*, a work which they had earlier performed, in April.

July also saw the publication of *Caniadau gan W. J. Gruffydd*, an undistinguished production which serves to indicate McCance's lack of real typographic skill. The Barcham Green Charles I paper which is a heavily tinted stone colour, though pleasant in itself, is unsympathetic to both the Perpetua type and to Hughes-Stanton's four pictorial wood engravings, and only a little less so to his shaded initials, though the black and Tyrian red ink harmonises well with it. The practice of underlining every title and of starting a new poem at the foot of the page is in questionable taste. Hughes-Stanton's otherwise agreeable title-page is marred by the inelegant shape of the letters used in the press device which appears to have been cut specially for the work. The binding has a wide greyish-brown buckram spine and paper covered sides. Hodgson devised the paper by means of using an ink roller to apply reddish-brown ink to the stone coloured background. The design for the special binding, in brown levant morocco, is by Hughes-Stanton. The decoration consists of a panel of twenty-four closely parallel gilt lines spreading across the cover, with bands of gilt lines at the top left of the upper cover and bottom right of the lower, so that the pattern on each cover is identical when reversed.

In September 1932, Hughes-Stanton's contract was terminated by mutual consent and replaced by a short term one, under the terms of which he was required to leave the Press by December when the printing of *The Revelation of Saint John* and the cutting of the blocks for *Four Poems by John Milton* were to be completed. The contract went on to stipulate that up to September 1933 Hughes-Stanton was to provide any blocks which might be required. In December the agreement was to be revoked and Hughes-Stanton agreed to stay at Gregynog until the following September.

The Press Board was not immune from a certain degree of philistinism, in one of its members at least. On 21 September, W. J. Burdon Evans wrote to T. J. concerning Hughes-Stanton's illustrations for *The Revelation of Saint John the Divine*. The implication was that he did not feel that the Press should proceed with printing.

> 'Some of [the illustrations] are very realistic, and I do not like them. Any dislike I may have does not arise from prudery but because I think them ugly. I cannot express any opinion worth having on works of art – I only know what I like.'

Happily, the printing was allowed to continue.

Minutes of a Press Board meeting held on 31 October reveal certain items of interest. Due to the continuing decline in the fine book trade it was decided to abandon plans for publishing Gilbert White's *Natural History of Selborne*, as the book was likely to be an expensive one. It was thought that the illustrations might well be used later for another book with less text! Gertrude Hermes had recently made a visit to Selborne from London, where she was now living, and the Board agreed to pay her four pounds to cover the cost of her expenses. As a further precautionary measure, it was resolved that the number of special bindings in future editions should be limited to fifteen copies. This number was thereafter maintained until the publication of the last four books which appeared 1938–1940. John Hugh Jones had claimed additional payment for reading from the stone, a task not undertaken by the other compositors, and Burdon Evans was instructed to enquire of local printers what an appropriate remuneration for this would be. Jones was later awarded four shillings per week for this extra responsibility.

The Singing Caravan: a Sufi Tale, by Sir Robert Vansittart, published in November, shows an improvement on McCance's previous productions. It is set in Perpetua type and exquisitely printed, again in black and tyrian red, a colour combination for which he had a strong predilection. The striking wood engraved frontispiece portrait is undeniably one of his more successful achievements at the Press. The asymmetric title-page is less satisfactory. The small capitals cramp the look of the title too much, and the justified setting of the imprint lends a lopsided look to the foot of the page. McCance's black initials, shaded with tyrian red, may appear a little over ornate to some eyes, yet here they do seem to be justified. In the colophon, Idwal Jones's name appears for the first time. The younger brother of Idris Jones, Idwal had recently joined the staff of the Press.

Local sheepskin, this time in a warm brown tone, was used again for the binding. McCance was responsible for designing the interesting and unusual special binding. Levant morocco was forsaken in favour of a burnt orange oasis, and the book has an Eastern type fore-edge flap, even though, unlike the bindings from which the idea was copied, boards are incorporated. The decoration consists of a complex arrangement of lines, enhanced by small black onlays, which extends over both covers, spine and flap. Fisher, who was responsible for translating McCance's rough charcoal sketch into a working drawing, claimed that the design was intended to convey a sense of music and light.

The only joint greeting card from the sisters for Christmas 1932 is a fairly simple one bearing the greeting message, printed in red and black, on grey silurian paper which encloses a leaf of eleven lines by Lascelles Abercrombie, with wood engraved initial letters by McCance. For the first time, Daisy Davies caused to be printed a separate, smaller card which is a simple fold bearing the greeting message on the inside recto and a wood engraving of the initials BC and a camp scene.[11]

1932 drew to a close with the Press having added three titles to its list of publications. Three earlier books – *The Plays of Euripides*, *Elia* and *Comus* had been selected by the First Edition Club for inclusion in their *Fifty Best Books of 1931* exhibition. Visitors to Gregynog during the year had included Christopher and Lettice Sandford, then in charge of the Golden Hours Press; David Bland and Vivian Ridler, two young men who came to camp in the grounds and who were later to contribute much to the field of printing; Lovat Dickson; J. Clement Davies, whose home was nearby at Meifod, and George Bernard Shaw, who stayed on 26 and 27 August with his wife Charlotte and wrote in the visitors' book,

'26–27 August – days of blessed rest – solo conversationalist, G. Bernard Shaw.'

For the sisters 1932 had been a year of interest and excitement, tinged with some anxiety. In June their brother, David, had been created Baron Davies of Llandinam. T. J. had been suffering from ill health for some time and, in late August, had been sent with his wife for a rest cure at Aix-les-Bains at Gwen Davies's expense. During the autumn, Sir Walford Davies, so often the instigator of schemes for Gregynog, had conceived the idea of holding a festival of poetry and music at the Hall the following year. The prospect of such a venture delighted the sisters, especially Gwen who immediately started to lay plans. On 12 October she wrote in great excitement to T. J.,

'We don't want lions necessarily but dreamers of dreams.'

Due to their interest in music their circle of acquaintance included several composers and performers, a number of whom had already visited Gregynog and participated in concerts held in the Music Room. As lovers of poetry and the owners of a press which had already printed several collections of verse, the sisters had also come into contact with a number of poets. The visual arts were to be excluded except for the art treasures already gracing the house. They felt that, for the time being at least, they had had a surfeit of artists! Despite their great interest in art, they had perhaps come to realise that they were unsuited, mentally and emotionally, to deal with the artistic temperament at close quarters.

Thus the year closed on a note of optimism. The festival, which was to be held annually for the next six years, was to bring much pleasure, not only to the Davies's but also to a somewhat elite circle of people from Montgomeryshire and further afield who were invited to attend. Extra work was to be created for the Press in the printing of festival programmes, in the drawing up of which Dora Herbert Jones was to play an important role.

NOTES

1. Hughes-Stanton insisted on trying out the various types of finishing tool for himself so as to get some idea of their limitations.
2. Conversation with the author in 1969.
3. Three of Hughes-Stanton's bookbinding designs are on books printed after he had left the Press.
4. Sir John Rhys, 1840–1915, Celtic scholar and sometime Principal of Jesus College, Oxford.
5. Letter from T. J. to McCance, dated 14 December 1931.
6. Conferences held over the years were many and various. Organisations most frequently holding meetings at Gregynog were: The Advisory Education Committee of the Welsh National Council of the League of Nations Union, The Distressed Areas of South Wales and Monmouthshire Conference on Social Reconstruction, The National Council of Music and The South Wales Guide Camp.
7. The marriage later ended in divorce.
8. Information given to the author by Mrs. Ursula Vaughan Williams.
9. Gertrude Hermes had received the Board's decision to terminate her contract in June.
10. Letter from Agnes Miller Parker to T. J., dated 19 July 1932.
11. The initials denote Bovington Camp, a camp site for deprived children in South Wales in which Daisy Davies had a special interest.

12

1933: A Successful Year

'I DO NOT THINK I HAVE EVER SEEN PRESSWORK of such uncanny perfection,' Bernard Newdigate wrote of Samuel Butler's *Erewhon*, published by the Press in February.[1] Typographically the book is not so successful, for the 10-point Baskerville type seems a little too small for sustained pleasurable reading. The title-page is a typographical horror in which the imprint is turned sideways and made to run vertically up from the foot of the page, a fact which did not go unnoticed by Joseph Thorp when reviewing the book for *The Architectural Review*.[2] The twenty-nine wood engravings by Hughes-Stanton which form chapter openings are aptly charming though unlike his familiar style. It is a great pity that these are sealed off from the underlying text by a heavy double rule, as this breaks the continuity of the page. Plain sage green sheepskin was used for the ordinary binding. The special binding, in brown polished levant morocco, designed by Hughes-Stanton, has been much admired, though possibly for the sheer virtuosity displayed in the tooling than for any merit residing in this very eccentric design. The titling appears too confined within the gilt ellipses which cross the spine. The bands of close gilt lines, carried across both covers and spine, end in peculiar stirrup-like ends which are disquieting in that they terminate too abruptly the otherwise faultless flow of the lines. In order to achieve the apparently jointless effect of the lines, Fisher filed down a gouge so that its edge became acutely rocker shaped which facilitated turning at the ends. Despite its rich effect, it is an odd design artistically and not to be compared with most of Hughes-Stanton's other binding designs.[3]

Gwen Davies continued to lean very heavily on T. J. Faced with the forthcoming departure of Hughes-Stanton in September and the possible resignation of McCance, her mind now turned on the problem of their replacement, and in February she asked T. J.,

'Could you keep your eyes and ears open and perhaps sound some of your

Opposite: Title-page to Erewhon.

EREWHON

BY SAMUEL BUTLER

THE WOOD-ENGRAVINGS BY
BLAIR HUGHES-STANTON

THE GREGYNOG PRESS MONTGOMERYSHIRE

MCMXXXII

bibliophiles whether it would not be possible to get an *educated* manager, young and with a sound knowledge of typography of the highest kind?'[4]

Her dependence on T. J.'s staunch friendship in a deeper and more personal sense is revealed in a letter written later the same month, thanking him for the gift of a book for her birthday.

'My Very Dear,
. . . It is such a joy to know that we can share so much together – the wonder and the hidden beauty of life, the stuff dreams are made of, the holy bread, the food unpriced: to know that it is still there just as it was when I was a girl, acutely, deeply conscious of it all, before life had rubbed the bloom off things.

And you have brought it back again – the old power of communicating loveliness to others – I had it out in France – And now it has returned once more because you believe in me – you understand – you care. For I would be a light that shines which does not flicker nor grow dim, that you may see the beauty of the shadows. And I would be a song in your heart when days are weary; and I would be strong wings beneath you, bearing you far above all that is sordid, or petty or mean, that you may see the land that is very far off.'[5]

This touching and highly emotional letter shows how deeply in accord the two friends were, in Gwen Davies's mind at least.

The Press Board met on 6 March and was much concerned about sales. They agreed to take back twenty of fifty books supplied to John and Edward Bumpus's bookshop in order to prevent their being sold at a reduced price.[6] Concern was also expressed at the lack of orders for special bindings. As the finishing of these was the unaided work of George Fisher, production was always at least two years behind publication. It was now felt that, in order to promote sales, the designs should be concurrent with the design of the book. A specimen copy could then be bound and photographed for inclusion in the prospectus. This eminently sensible proposal was never put into action. Frequent changes in designers during the ensuing years no doubt rendered it impracticable. It was at this meeting that the Board vetted McCance's design for the special binding of *The Singing Caravan*. It was felt that the intricate construction and design might take a disproportionate amount of George Fisher's time – too much for the market as it was. It was decided that the fore-edge flap be omitted except for a few copies bound for the Davies's and their family, but, when the time came for binding, this economy measure was abandoned, and all twenty-five copies were bound complete with flap. One or two of the special bindings were already proving something of an embarrassment. It had

Opposite: Page from Erewhon, *with engraving by Blair Hughes-Stanton.*

CHAPTER TWENTY-NINE

CONCLUSION

The ship was the *Principe Umberto*, bound from Callao to Genoa; she had carried a number of emigrants to Rio, had gone thence to Callao, where she had taken in a cargo of guano, and was now on her way home. The captain was a certain Giovanni Gianni, a native of Sestri; he has kindly allowed me to refer to him in case the truth of my story should be disputed; but I grieve to say that I suffered him to mislead himself in some important particulars. I should add that when we were picked up we were a thousand miles from land.

As soon as we were on board, the captain began questioning us about the siege of Paris, from which city he had assumed that we must have come, notwithstanding our immense distance from Europe. As may be supposed, I had not heard a syllable about the war between France and been the custom to present one of these gratis to each author or editor as part or token payment. This had not happened in the case of Gilbert Murray, but as several specially bound copies of *The Plays of Euripides* still remained unsold, it was now agreed that he should receive one after all.

The matter of finding replacements for Hughes-Stanton and for McCance, who had now announced his intention of leaving the Press in September also, was under active consideration. Dora Herbert Jones put forward the names of David Bland and Vivian Ridler who had visited the Press the previous summer, but it was felt that they were too young and as yet untried. Vincent G. Stuart, who was then running his own private press at Brockweir, near Tintern, wrote enquiring about the post of controller and enclosed a curriculum vitae. W. J. Burdon Evans was requested by T. J. to make it unequivocally clear to Blair Hughes-Stanton that his contract would not be renewed beyond September but that he was to feel free in the future to suggest for publication any title which he felt drawn to illustrate. T. J. urged a tactful approach in case the Press should want to use Hughes-Stanton's services again when the market had absorbed his current output!

In common with the generality of wealthy people the Davies's, fearing that they would be taken advantage of and thus always nervous of those around them, frequently entertained a parsimonious attitude towards their staff. Nevertheless, they, and Gwen Davies in particular, were often generous in other respects. In April T. J. was sent to a clinic in Ruthin, Dora Herbert Jones to Llandrindod Wells and W. J. Burdon Evans to Bath, all at her expense, to undergo treatment for various ills. Pecuniary aid was often dispensed by stealth as frequently evidenced by her correspondence. It is not known how the sisters first made the acquaintance of the poet Lascelles Abercrombie though no doubt they were introduced by T. J. In May Gwen wrote to T. J. saying that she had it on good authority that the Abercrombies were delightful people but 'dying by inches' from overwork and poverty. This had given her an idea. Would it not be splendid if she, aided by friends, were to collect sufficient funds to send the Abercrombies and others in a similar plight, whom she described as 'a bunch of needy intellectuals' for a rest cure and treatment at Ruthin during the summer vacation?[7]

For some time now the *cognoscenti* had been awaiting the appearance of that superb folio *The Revelation of Saint John the Divine* which was finally published on 5 April. The illustrations, which, as we have seen,

Opposite: Title-page to The Revelation of Saint John the Divine, *printed in Tyrian red and black, with engraving by Blair Hughes-Stanton.* (*Reduced.*)

THE REVELATION OF SAINT JOHN THE DIVINE

THE
GREGYNOG PRESS
MCMXXXII

had caused disquiet in the breast of at least one member of the Press Board, show Blair Hughes-Stanton at the full height of his powers as a wood engraver. These engravings, in which the strange elongated figures exude a strange power and even menace, emphasise the artist's complete technical mastery of the medium. They attracted considerable attention in an *Observer* review which appeared on 18 June:

'The Gregynog and the Golden Cockerel are perhaps the only two presses that can be mentioned in the same breath with the Ashendene . . . [Hughes-Stanton's] reputation as an illustrator has been steadily and deservedly mounting. He has done particularly good work in the past for the Cresset Press . . ., but there can be no doubt that he has here surpassed himself. There is in the wild sanity of this astounding book something very close to the texture of Mr. Stanton's own mind. He commands and controls a savage dexterity that flows from an ardent imagination. No limits are set to that imagination by the Book of John of Patmos. He has taken the stars to juggle with, and all heaven and earth for his stage. His theme is as immeasurable as his audacity . . . Indeed the artist, untroubled by historical or metaphysical speculation, has been able to hold up an unobscured torch which has helped at least one reader to see the dark revelation anew.'[8]

This view was not universally shared however. Bernard Newdigate, writing in *The London Mercury*, opined,

'. . . yet many will feel that he has failed in his interpretation of the apocalyptic vision. His treatment will offend many of those who treasure the tradition of Christian iconography.'[9]

Surely this unique book is beyond carping of this sort? It may not be perfect in all respects, but what a stimulating visual experience it is simply to turn the pages, each one providing a new feast for the eye.

The illustrations are further enhanced by the text, set in Bembo type with Perpetua titling, and printed in black and Tyrian red on heavy Japanese vellum. Hughes-Stanton also provided the superb title-page which incorporates an ingenious press device with the GG initials interlaced, back to back, and superimposed on a cross. The binding is in a deep red Hermitage calf, titled in blind on the front cover and spine. The design of the special binding is one of Hughes-Stanton's most conspicuously successful achievements in this field. Bound in cream coloured unpolished levant morocco, each cover bears a crushed panel, the larger one, on the upper cover, carrying symbolic ornaments, including a black onlaid cross with fine shading and gilt lettering, which is contrasted with a dark brick-red onlay, lettered and shaded in

Opposite: Page from The Revelation of Saint John the Divine, *printed in Tyrian red and black, with engraving by Blair Hughes-Stanton.* (*Reduced.*)

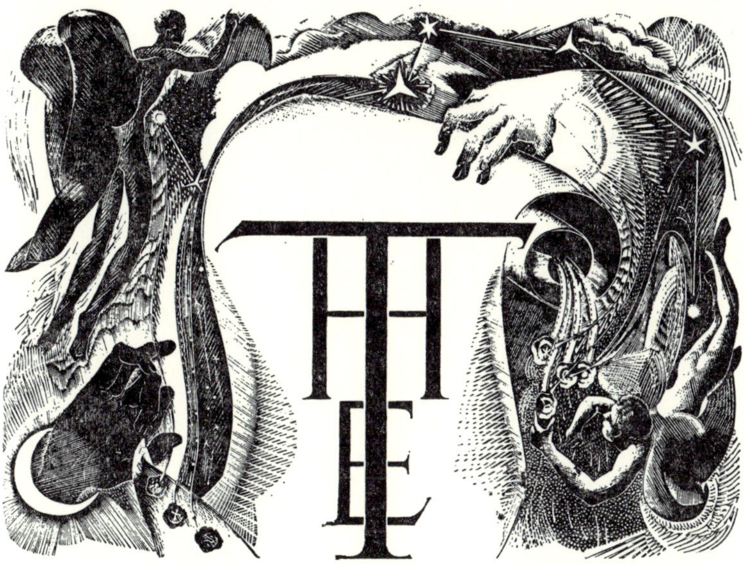

REVELATION
OF JESUS CHRIST,
WHICH GOD GAVE UNTO HIM

TO shew unto his servants things which must shortly come to pass; and he sent and signified it by his angel unto his servant John: who bare record of the word of God, and of the testimony of Jesus Christ, and of all things that he saw. Blessed is he that readeth, and they that hear the words of this prophecy, and keep those things which are written therein: for the time is at hand.

JOHN

TO THE SEVEN CHURCHES WHICH ARE IN ASIA: Grace be unto you, and peace, from him which is, and which was, and which is to come; and from the seven spirits which are before his throne; and from Jesus Christ, who is the faithful witness, and the first begotten of the dead, and the prince of the kings of the earth. Unto him that loved us, and washed us from our sins in his own blood, and hath made us kings and priests unto God and his Father; to him be glory and dominion for ever and ever. Amen. ¶ Behold, he cometh with clouds; and every eye shall see him, and they also which pierced him:

blind. The title appears on the upper cover and the imprint on the lower.

Work in connection with the forthcoming festival of music and poetry took precedence in the summer as several pieces of printing were required. These comprised an invitation and separate reply card, an order of service, the festival programme itself, which included libretti, and separate double-fold programmes for each of the four days, intended for those guests not invited to attend the whole festival. Plans were made to accommodate the participating artists and chief guests in the Hall itself while lesser mortals were to be accommodated at the Buck Hotel, a small country inn, situated in the village of Caersws, a few miles from Gregynog.[10]

The day of the great festival dawned at last. It opened with a concert of choral and instrumental music and songs at eight o'clock on the evening of Friday, 23 June. The company assembled in the Music Room. Round the hessian-clad walls, which afforded excellent acoustic qualities, were the lovely French and English paintings, including four Monets in a row. As the audience filed in, their gaze met the calm, authoritative eyes of Christ, looking out from a large painting of the Via Crucis by the school of El Greco which stood on a screen at the front of the stage. The picture obscured organist, conductor and a considerable portion of the choir. As befitted the quiet, unobtrusive tenor of life at Gregynog, personalities were kept in low key: the performance not the performers mattered. The choir, conducted by Adrian Boult, opened the concert with Hubert Parry's *Ode at Solemn Musick*. After this, they sang a group of English madrigals and Gustav Holst's setting of Keats's *Ode on a Grecian Urn*. The remainder of the programme consisted of songs sung by Megan Foster and instrumental pieces played by Raymond Jeremy, Sybil Eaton and Walford Davies who was musical director in chief of the Festival. Poems were read by Ellen Bowick, a teacher of elocution who had tutored the Davies sisters.

On Saturday, a brief concert of chamber music was performed after tea, followed by the main concert of the day at eight o'clock. The first half was devoted to readings of modern poetry by Lascelles Abercrombie and, after the interval, Elsie Suddaby sang music by Armstrong Gibbs, Herbert Howells and Bach. A group of strings played Brahms's *Sextet for Strings in B Flat. Opus 18*. Two well known names featured in the sextet, the two violinists, Jelly d'Aranyi and her brother-in-law, Alexander Fachiri.

Sunday was a pretty full day. It began with a service at ten-thirty.

Opposite: Engraving by Blair Hughes-Stanton, from The Revelation of Saint John the Divine.

A brief concert followed in the early evening entitled *Wales, her History and Folk Music* in which Dora Herbert Jones sang the folksongs. After dinner, came the main concert which consisted mainly of choral works including Walford Davies's *Reverie*, newly composed for the Festival.

The final concert was held on Monday evening. Once again it consisted of choral works, songs sung by Elsie Suddaby and instrumental pieces, during which Lascelles Abercrombie read his own *Ceremonial Ode*. The Festival drew to a close with the choir singing the Bach chorale *Zion Hears Her Watchmen's Voices*.

It was over, the event which the sisters, ably aided and abetted by Dora Herbert Jones, had been planning for the last eight months. The general consensus of opinion was that it had been an unqualified success. Artists and audience alike had enjoyed their weekend. The audience, however, did have one or two gaps. The Shaws had been invited to attend, but G.B.S., clearly not keen to come, had enquired whether Gregynog had a padded room in which he could get away from Holst's music!

At this, as at all subsequent festivals, the audience were invited to make donations for a charity. The one selected on this occasion was Coleg Harlech, an enterprise closely connected with T. J., and which, at that time, was serving mainly as a training centre for the unemployed. The opening paragraphs to the foreword in the festival programme make interesting reading, as they clearly come from the pen of Gwen Davies:

> 'In these days of unprecedented difficulty and disillusionment, when the very fabric of our civilisation is rent and torn, we are compelled to return again to the unfailing sources of inspiration and delight. Poetry and music are no longer the luxury of the few but the necessity of the many.
>
> We have conquered the forces of nature, harnessed the waters for our profit, and unlocked the secrets of the air, but we have failed to apprehend the unseen beauty surrounding us, pressing in on every side, enfolding us with loveliness, could we but know it.
>
> Thousands upon thousands of our fellow human beings are dragging out a dark and desolate existence; exhausted and in despair they stand at the corners of the streets, for no man hath need of them. A bewildered Government doles out to them the pittance which keeps them alive, but their minds and spirits are starved, for man doth not live by bread alone. They are bone of our bone and flesh of our flesh, their powers and potentialities are the same as ours, the same unseen beauty is theirs, could we but show it to them.'

Again one hears not only the voice of human sympathy but also the echo of a guilt which no amount of giving could assuage. A note of thanks from the sisters to all donors was printed at the Press later in July which announced that the sum of £112 had been sent to the treasurer of Coleg Harlech in response to the appeal.

The festival momentarily eclipsed events taking place in the Press, but, in fact, O. M. Edward's *Clych Atgof* had been published just prior to the festival. There is a marked contrast between this and the last book, *The Revelation of Saint John the Divine*. This is very much McCance's book. The text is set in Perpetua and printed on Batchelor's special Gregynog paper and enhanced with a frontispiece and eight small wood engraved chapter openings, all by McCance. The openings, printed in reddish brown, are pleasing in themselves if a little mechanical, but their effect is spoilt by being placed too close to the text and by being overprinted with black capitals, set on a brown background, which line up neither with the engravings nor with the text lines. The title-page is an unsatisfactory mixture of symmetric and asymmetric typography, not improved by the addition of a weak, elongated press device. Natural colour Welsh sheepskin was used for the ordinary binding. Author, title, imprint, device and date are incorporated in a block extending over both covers and spine which was printed on the printing press by Hodgson, using reddish brown ink. The special binding, on the first fifteen copies, which is in a polished dark brown levant morocco, was designed by George Fisher, but the design is sadly uninspiring. Two bands of gilt lines, terminating in lozenges, stretch across both covers, while a perpendicular arrangement of the same design decorates the spine.

The minutes of a Press Board meeting, held in July at the Davies's London flat, reveal continuing anxiety about the state of the market. It was resolved to restrict the number of copies of the forthcoming *Lamentations of Jeremiah* to one hundred and seventy-five copies. By the time the instruction reached the Press, it was already too late: the printing of the edition was virtually complete. McCance agreed to leave one hundred copies in sheets for the time being, but, in the end, this proved unnecessary, and all two hundred and fifty copies were bound. Burdon Evans proposed that, in future, the Press should print fewer books and limit them to very special items. *The Welsh Outlook* had reached a very low ebb, and its demise seemed inevitable.[11] McCance came up with the suggestion that its publishers and the Gregynog Press might combine to produce a quarterly magazine to serve as the cultural organ of the voluntary service movement, and that Coleg Harlech might possibly be invited to co-operate. He pointed out that such a venture would have the dual advantage of keeping the Press staff fully occupied in slack periods. Burdon Evans was in favour of the suggestion, but T. J. was clearly not enamoured of the idea, and it died an early death.

Early in 1933, W. H. Davies began to grow agitated concerning the non-appearance of his collection of poems which had been lying fallow now for the last two years. It was an embarrassing situation, for Gwen Davies was understandably reluctant to ask McCance and Hughes-Stanton to work on another design for the book, having rejected their earlier version. T. J. found a possible solution to the dilemma in suggesting that Gertrude Hermes be asked to prepare another set of illustrations.

Burdon Evans was sceptical of the plan to get her to redesign *The Lovers' Song-Book* and suggested that she be asked to redesign *parts* of it, for a fee not exceeding fifty pounds. In a letter to Dora Herbert Jones addressed from Adam's Grammar School, Newport, Salop, where she was now teaching, Gertrude Hermes declared her preference for starting the book from scratch.

'. . . having been given no opportunity by the Press, my reputation as a productive engraver is nil, and I have had to turn my activities in other directions and am now engaged in decorative work and portrait busts.'[12]

Her request was not granted, largely on economic grounds.

A new set of wood engravings was submitted for the Board's approval by the artist late in September. Alas, these met with a response similar to that which had greeted Hughes-Stanton's earlier engravings for the same book, and they were never published. Dora Herbert Jones commented to T. J.,

'. . . with these black masses of engraving and with only a tiny fragment of print on the other side, the book will look more like a memorial roll rather than a lovers' song book!'[13]

So ended Gertrude Hermes's tenuous and unhappy association with the Press. Looking back on her long and highly successful career as wood engraver and sculptor, one cannot but regret that Gregynog never succeeded in making use of her very considerable talent, and reflect that had times and circumstances been other than they were, for her and for the Press, her lovely work might have further enhanced Gregynog's reputation.

Since McCance had intimated his plan to leave the Press in September, anxiety was now expressed concerning the lack of progress on work for *The Book of Job*, a book which McCance himself had suggested for publication and for which he had offered to provide the illustrations. Though the text had been set by July, only five of the projected fifteen blocks had so far been cut, and the initial letters were not yet begun. It now became apparent that McCance's interest in the book was waning fast, and he went so far as to suggest that the initials should be designed and cut elsewhere. Nearly two months before he left the Press, he received a letter from W. J. Burdon Evans which firmly announced that should all the blocks not be cut by 30 October, the Press Board would not sanction any further payment for them. He went on to suggest that, as time was short, McCance's wife might care to co-operate with him. And there the matter rested, for the blocks were never completed, and the type was eventually distributed early in 1935 when it was needed for other printing, an event which grieved Gwen Davies sorely when she thought of the money wasted on its composition!

On 25 July, and again in August, an advertisement appeared in the press for a new controller. T. J. took care to preserve a list of the most likely applicants. Vincent G. Stuart had been vetted by Stanley Morison at T. J.'s request but is not among those listed. The list included Donald Attwater, who had applied previously, after the departure of Maynard; Vincent M. Beach, then at Cambridge University Press, and R. P. Ross Williamson, who had applied on the recommendation of Brooke Crutchley. Williamson was also recommended by Stanley Morison who declared the applicant's lack of experience to be of small account when set against his great enthusiasm. Two others were Harry Carter, then at the Kynoch Press, and Reynolds Stone, who was then in charge of the machine room at the Wessex Press in Taunton. Stone visited the Press in early August when W. J. Burdon Evans found him sadly wanting in the practical virtues! Gwen Davies thought this of little account, being of the opinion that Burdon Evans himself could supply the necessary business expertise:

'The artistic qualities (good taste, sound artistic flare [*sic*] and education) are *essential* to us: someone who can *produce* a fine book is what we need – there are plenty of the other kind to be had, and our Press can in no way be considered a commercial proposition.'[14]

In writing to recommend Ross Williamson, Stanley Morison, who had made his one and only visit to Gregynog on 25 August, took it upon himself to make a few suggestions concerning the future policy of the Press. He pointed out that its interests would not be best served by frequent changes of staff, and that the Press Board should seek to stabilise its position. In his opinion, the controller should be in charge of publishing but that he need be no more than sympathetic to the aims of typographer and artist. T. J. and Burdon Evans should themselves take more account of style, and that, to this end, they should train themselves from books. The success of the Press would not then depend on regular importation of artists and other experts from outside. He went on to criticise the style of colophon adopted at Gregynog, saying that much of the material, including names, could be relegated to some future history of the Press. Finally, and not surprisingly, he considered that they should cut down on the number of illustrated books produced.[15] It seems surprising, in view of the condemnatory tone of Morison's dealings with the Press, that T. J. should have continued consulting him. He seems not to have taken the time or the trouble to discover anything about the Davies's aims or surely he would not have made the recommendation he did regarding illustrated books. His protégé, Williamson, was invited to visit the Press early in September but was not appointed.

John Johnson, printer to the Oxford University Press, to whom T. J. often turned also for advice, was much more understanding in his approach. Here was a man who had no desire to be lionised but merely to be helpful. At the end of August he had written to T. J. concerning Loyd Haberly, a young American from a poor background who, by his own efforts, had managed to get to Harvard, where he had eventually become a lecturer in international law. As recipient of a Rhodes scholarship, he had come to England where he had taken a law degree at Trinity College, Oxford. Since then he had occupied himself in writing, printing and bookbinding.[16] He was now running his own private press, the Seven Acres Press, in a building which he had built himself, even to trimming his own stone. The Press was in the village of Long Crendon, Buckinghamshire, adjoining the home of Mrs Arthur Durnford and her ward, Miss Agatha Walker.[17] Johnson also mentioned that the University Press had already published a book of verse by Haberly which the printer considered unprecedented in view of Haberly's extreme youth. Haberly, it appeared, was also interested in archaeology and had recently taken it upon himself to excavate the site of Notley Abbey in the vicinity of Long Crendon.

Johnson had no doubt that here was the man for the job, even though he admitted that Haberly would probably need three months training at O.U.P. first.

'If ever a man had the right spirit, Haberly has it . . . and is that blend of dreamer and practical man which it is so difficult to find in life. It is a blend which you more often meet, if I may say so, in your race than mine.'

The letter went on to confess a personal affection and admiration for Haberly whose printing he praised highly. Not only did Haberly do everything he attempted well but also got on well with people and stimulated those around him.[18]

Nothing could have been calculated to find greater favour in Gwen Davies's eyes. Was not 'a blend of dreamer and practical man' exactly what was wanted for Gregynog? Haberly was invited to Gregynog within a week and was offered the job soon afterwards. The young man was clearly not over impressed with what he saw for he wrote rejecting the offer. He felt that he had too many other interests which lay at too great a remove from the depths of mid-Wales. So great was the desire of the directors to secure his services, however, that T. J. wrote to John Johnson, asking him to put to Haberly the offer of a part-time appointment. It was suggested the he should spend one week of every month at Gregynog for which he would be paid a salary of £200 per annum. To this Haberly eventually agreed.

In view of the part-time nature of the forthcoming appointment, T. J. recommended that a full-time works manager be appointed in addition to the controller, and several candidates were considered. These included Vincent Beach of the Cambridge University Press and E. M. Cheetham, who was then printing for Fullers the confectioners. Both of these were recommended by Walter Lewis. J. G. Wilson, manager of John and Edward Bumpus's Bookshop, recommended H. Charles Tomlinson, son of the novelist, and David Bland wrote offering his services. Dora Herbert Jones, whom it had already been agreed should take charge of the Press during the interim, was of the opinion that they could cope very well without a manager. The Board eventually decided not to make such an appointment, a decision which perhaps, in the light of later events, it came to regret.

While negotiations over appointing a controller engaged the attention of the directors that summer and autumn, the Press continued its printing activities undisturbed. Several items of ephemera were prepared, and three books, all published after McCance and Hughes-Stanton had left the Press, were also completed.

Four Poems by John Milton, published in October 1933, had been finished as long ago as 5 January. It is a monochrome work, set in Perpetua and printed on Japanese vellum. Once again the setting suffers from the format imposed by Hughes-Stanton's eleven magnificent figure engravings, in that the type area appears like an over-long ribbon, though in general it is a pleasing book. There is no doubt here that the poems selected are merely a vehicle for displaying the illustrations.

The ordinary binding is one of the most successful to come from the Press. It is in a rich reddish-brown Hermitage calf, decorated in blind on the front cover with a block derived from the engraving of Euphrosyne which appears on the title-page. Nearly a year after the book was published, Hughes-Stanton was asked to provide a design for the special binding. The design eventually submitted had to be amended to fit in with the sewing which had already been carried out. The book is covered in a dun coloured unpolished levant morocco. Both covers are decorated similarly with perpendicular crushed panels each containing two panels of lines, the top one a blind lattice and the lower one formed of horizontal gilt lines. The two motifs are repeated in the second and fourth spine panels. Only fourteen copies were bound to this design: Number 15 has a unique binding, in rose pink levant morocco, with a design by George Fisher in which an all-over pattern of perpendicular lines terminates in lozenges onlaid in a contrasting colour and outlined in blind.

XXI Welsh Gypsy Folk-Tales, collected by John Sampson and edited by Dora Yates, was completed in June and published in November. This demonstrates a sympathetic use of Bembo type on Portal's hand-made paper. The title-page and seven chapter openings are by Agnes Miller Parker. These are as spirited in conception and design as they are delicate in execution and demonstrate the artist's skill to perfection. Though, once again, the illustrations are a little closely crowded by type, nonetheless, it is a lovely book. Dora Yates who edited the book also acted as Sampson's literary executor and kept a watchful eye on all stages of production during its preparation. From reading the correspondence between her and McCance it is clear that the appearance of the text pages was very much improved due to several pertinent comments she made.[19]

Mustard yellow was chosen for both bindings. This again was at

Opposite: Page from XXI Welsh Gypsy Folk Tales,
with engraving by Agnes Miller Parker. (*Slightly reduced.*)

THE GREEN MAN OF NO MAN'S LAND.

THERE WAS A MILL, A YOUNG MAN, and his maidservant. Now this man was a great gambler. No one could beat him: he beat everybody. ℭ There came a gentleman to him, and he walked straight into the miller's room. The two had a word together. "Wilt thou play?" asked the gentleman of the young man: the cards were on the table. "Yes," quoth the young man. ℭ Now the two are at their game. The miller beat the gentleman. The gentleman asked Jack: "What wouldst thou have?" "I would have a castle yonder." As soon as he said the word, there was the castle. ℭ The

Dora Yates's instigation in that she specifically stated that it must be bound in a 'gipsy colour, either red or yellow'. The Welsh sheepskin binding of the ordinary edition is blocked on the upper cover with decorative gilt titling. The design of the levant morocco cover of the special binding is by George Fisher. Broad gilt bands stretch across top and bottom, the centre being occupied with vertical lines of lozenges, onlaid in red and black, which are joined by gilt lines. Bands of vertical gilt lines occupy the top and bottom spine panels. On the turn-ins oblique gilt lines are tooled at top and bottom, while more onlaid lozenges appear down the fore-edge.

The Lamentations of Jeremiah, printed from the text of Scrivener's *Paragraph Bible*, was completed in August and published eleven months later. This is a superb folio, set in Baskerville italic and printed on heavy Japanese vellum. Continuous setting in an italic type is rarely satisfactory, but this is an exception. There is a wood engraved title-page of astonishing wizardry by Hughes-Stanton which incorporates a unique GG monogram device, superimposed on a cross, similar in some respects to that employed on the title-page of *The Revelation of Saint John the Divine*, but in this case the G's are imposed one above the other and not back-to-back. There are five full-page engraved chapter openings and other engravings also by Hughes-Stanton which consist mainly of restless figures, portrayed nude or in diaphanous clothing. Bernard Newdigate, in his review in *The London Mercury*, was full of admiration for the skill displayed in these engravings, as well he might have been, but commented,

'. . . but neither these nor any other engraved pictures can convey the sense of utter desolation borne by the bible text.'[20]

The illustrations, the setting of the text and the large wood engraved initials and running heads, printed in blue, all conspire to render this one of the most spectacular books to come from this or any other press.

One hundred and twenty-five copies were bound in nocturnal blue Hermitage calf and one hundred and ten copies in blue oasis, with lettering blocked in blind on the upper cover and the spine. The choice of blind for the blocking was a mistake, as it does not show up at all well on the dark blue covers. The special binding, designed by Hughes-Stanton, for the fifteen remaining copies in black levant morocco is much more successful. Both covers here are identical. Each has an onlaid disc of dark blue, shaded with gilt and blind lines, from which

Opposite: Engraving by Blair Hughes-Stanton
from The Lamentations of Jeremiah.

hangs a symbolic avenging sword, over onlaid white squares of diminishing size, outlined in gilt. The bases of the squares are linked by a diminishing arrangement of lines, while a fine gilt line connects them with the upper part of the design. It is a design of great complexity and superbly executed by George Fisher.

Blair Hughes-Stanton and the McCances finally left the Press in September 1933, and there is no doubt that the directors were pleased to see them go. Neither side had found the three years of their stay particularly easy. There had been constant friction and clashes of personality within the Press itself. Furthermore, much of their work had been the cause of unhappy speculation on the part of the Board. The directors had never been too pleased with McCance as a typographer, perhaps with some justification. Though Hughes-Stanton's virtuosity had excited much admiration, his style was really a little too *avant garde* for their taste: they had been much more content with the more staid and more easily assimilable art of Maynard and Bray. In general their work was undervalued by their masters, and the work of Gertrude Hermes to the point of oblivion.

Whatever their faults and personal shortcomings, this group of artists left behind them a legacy of illustrated books which were the envy of other presses.[21] It must, however, be said that much of the success of these was due to Herbert Hodgson's incomparable skill as a printer. Hughes-Stanton's blocks have never been better printed, as the artist is the first to acknowledge.

Never again was Gregynog to have a resident controller or artist. If the Press Board could have foreseen the problems which were to follow, it would perhaps not have been quite so anxious to usher in a new regime.

Following the departure of McCance and Hughes-Stanton, work continued in the Press under the watchful eye of Dora Herbert Jones. At the request of Mrs N. P. Campbell of Tunbridge Wells, Gwen Davies had agreed to the printing, at her expense, of an account of the social service movement initiated by Beatrice Hankey. This eight-page pamphlet, entitled *The Blue Pilgrims*, was printed in an edition of one thousand copies in October. Due to the pressure of work, the text of an order of service for the South Wales Guide Camp, held at Gregynog in December, was once more sent out to Western Typesetting Services for composition.

Having despaired of obtaining a satisfactory design for an illustrated edition of W. H. Davies's *The Lovers' Song-Book*, the Board decided to prepare a straightforward unillustrated edition in time for the Christmas market. This was designed by Dora Herbert Jones, assisted by John Hugh Jones. T. J. requested that this information be given in the book's colophon, but Dora Herbert Jones was adamant that its design should remain anonymous, and the colophon merely states that the printing of the book was completed on 11 November. Not surprisingly, this is one of the least interesting of the Gregynog books. It was intended to be a complementary volume to W. H. Davies's *Selected Poems*, published in 1928, and was thus a straightforward setting in Baskerville type on heavy Japanese vellum with ruled pages. Most of the poems are very short and seem to be set too high within the ruled frames. As an economy measure, materials left over from other books were used, so that copies are variously ruled in either blue or green, with coloured initials to match. The ordinary bindings are also found in variants of blue or green Cockerell marbled paper sides with toning cloth spines. The first fifteen copies, plus four copies for the Davies's and friends, were bound in a light green levant morocco. Fisher was responsible for the design which is simple and somewhat stereotyped. Tooled foliage springs in pyramids from the base of each cover and the spine, and is combined with small lozenges. Fisher in later years said,

> 'All-over ornament was not in keeping. I felt the richness and quietness were appropriate.'[22]

The whole book compares very unfavourably with the illustrated version prepared by McCance and Hughes-Stanton in 1930 which was rejected by the Press Board.

T. J. had prepared a collection of sixteen of his speeches and articles which the Press printed in an edition of one hundred and seventy-five copies for private circulation in December. It was dedicated 'To Gwendoline and Margaret Davies, founders of Gregynog' and included a reprint of the article on the Press which had appeared in *The Western Mail* in December 1925. The volume was originally intended to be titled *Footprints*; a binder's brass for this title was actually made, but T. J. later changed it to *A Theme With Variations*. This has a straightforward setting in Perpetua type and is printed on a variety of handmade papers. (T. J. gave as the justification for producing the volume that paper left over from earlier books could be used up.) It is bound in the brick red cloth which was later used for the binding of *The Story of a Red-deer,* with a monogram of the author's initials blocked in gilt on the upper cover.

Loyd Haberly had not made an appearance at Gregynog since his appointment, but, in October, had sent proofs of some of his own poems which he thought the Press might like to print to mark, as he put it, 'the beginning of the third dynasty'. His presence at a meeting of the Press Board, held on 28 November, is recorded in the minutes, though it is likely that the meeting was held in London.[23] At the meeting, Haberly suggested for publication an edition of Robert Bridges's *Eros and Psyche*. Haberly had been acquainted with Bridges before his death in 1930, and had sought and obtained the approval of Mrs Bridges. At this point, Haberly was in favour of setting the work in Bruce Rogers's Centaur type but announced that he was investigating the possibility of designing special types for Gregynog's use, one perhaps for poetry and one for prose!

Haberly was not to take up his position formally until January 1934, but the first piece designed by him was in fact the Christmas card for 1933. It is a fairly simple affair and not at all typical of his later work at Gregynog. The greeting from the sisters is printed on the grey silurian covers which enclose a broadside of lines beginning, 'The future lies large and splendid before us . . .' opening with a large wood engraved T.

It had been a busy and interesting year all round. Six books had been printed and six published. (Only four of these were printed *and* published in 1933, as publication was frequently delayed by several months.) During the autumn a set of the Gregynog special bindings was exhibited at the First Edition Club where they attracted a lot of attention. The first festival of music and poetry had brought a number of distinguished performers and interesting guests to Gregynog. It had been all that the sisters had hoped for, and seems to have been attended by fewer frustrations and disappointments than some of the later festivals. Other well known visitors that year included Stanley Morison, the Shaws and John and Susan Buchan (Lord and Lady Tweedsmuir) who, accompanied by the Fairfax-Lucys and Mostyn Lewis, had visited the Press on 30 August. Relations between the two sisters and their brother, David, were much improved, and he spent two days with them at Gregynog in June. In July Gwen and Daisy were greatly alarmed when he collapsed, due to overwork, in a Cardiff street, but by the end of the year he had recovered his old vigour once more.

NOTES

1. Bernard Newdigate. Review in *The London Mercury*, Vol. XXVI, April, 1933.
2. *Junius* (pseud. of Joseph Thorp). Review in *The Architectural Review*, April, 1933.
3. Hughes-Stanton's design was intended to be executed partly in blind.
4. Letter from Gwen Davies to T. J., dated 18 February 1933.
5. Letter from Gwen Davies to T. J., dated 23 February 1933.
6. The title, or titles, to which this refers is not revealed.
7. Letter from Gwen Davies to T. J., dated 2 May 1933.
8. The review is unsigned but would appear to have come from the pen of J. L. Garvin, the editor.
9. Review by Bernard Newdigate in *The London Mercury*, Vol. XXVIII, June, 1933.
10. While the sisters paid jointly for the entertainment of guests to the summer festivals, the artists' fees were provided out of Gwen Davies's private account.
11. It ceased publication very soon afterwards.
12. Letter from Gertrude Hermes to Dora Herbert Jones, dated 11 July 1933.
13. Letter from Dora Herbert Jones to T. J., dated 12 October 1933.
14. Letter from Gwen Davies to T. J., dated 13 August 1933.
15. Information taken from a letter from Stanley Morison to T. J., dated 1 September 1933.
16. He had learnt the rudiments of printing, engraving and bookbinding at evening classes held at the Central School of Arts and Crafts and had received further coaching from Emery Walker.
17. Agatha Walker, a cousin of Joan Evans, an artist who modelled exquisite miniature wax portrait figures which were exhibited more than once at the Royal Academy.
18. Letter from John Johnson to T. J., dated 31 August 1933.
19. The correspondence is owned by Mrs. Margaret McCance, the controller's widow.
20. Review by Bernard Newdigate in *The London Mercury*, Vol. XXX, September 1934.
21. Blair Hughes-Stanton alone cut more than two hundred blocks while at the Press.
22. Conversation with the author in 1969.
23. The Board meetings' minutes do not always reveal where meetings were held.

13
Loyd Haberly

AS AN ADDITIONAL EMOLUMENT FOR TAKING charge of the Press from September 1933 until January 1934, the Press Board awarded Dora Herbert Jones the sum of fifty pounds. John Hugh Jones also received ten pounds extra responsibility allowance. As noted earlier, Mrs Herbert Jones's position in the Press was not an altogether enviable one. The situation, during McCance's controllership, had become more difficult, largely due to a mutual lack of sympathy. On her side, this sprang from a lack of confidence in McCance's typographical ability, and, on his side, from a misplaced lack of trust in her integrity, since he appears to have suspected her of acting as a watch dog for the owners. With Haberly's appointment, the role of go-between betwixt the staff working in the Press and an absentee controller seemed to have been thrust upon her. Furthermore, she felt her position as secretary to have been undermined. Part of the problem stemmed from the fact that she had to work more closely with W. J. Burdon Evans who had now assumed the mantle of general overseer of Press affairs. Evans, a man incapable of appreciating the artistic aims of the enterprise and not used to dealing with women, was clearly antipathetic to her frustrations.

For some obscure reason, Dora Herbert Jones began to suspect that the sisters wished for her resignation. Perhaps this was true of Daisy who appears to have been somewhat jealous of the secretary. Certainly she had been against Dora being left in charge of the Press for the last few months. It was not true, however, of her sister, as a letter from Gwen Davies sent to T. J. on 25 January clearly demonstrates:

'Dora is *vital* to me in many ways as our minds work together . . . She is a rapid thinker. You don't have to cross your T's or dot your I's . . . Neither Daisy nor Miss Blaker have any conception of the nervous energy, the electric current that has to run through things to make Gregynog "alive" . . . Daisy is a brick and does splendid work, but a brick is inclined to be, and must be, unbending and rigid.'

In January, Dora announced that she intended to apply for the post of warden in one of the halls of residence at University College Cardiff. One suspects that this was a gesture of defiance, as she did not in the least wish to leave Gregynog, a fact only too apparent in a letter sent to T. J. on 31 January:

'I think you know what it means to me to leave my beloved printing, the services and the choir . . . Miss Gwen told me on Saturday that *she* did not want me to go, but she has done nothing beyond that to show that it was worthwhile doing something to keep me!'

This interchange between the two women took place during a train journey back to Gregynog from London where a Press Board meeting had been held on 24 January at which Loyd Haberly was also present. The Secretary reported on items printed since September. These comprised seven hundred copies of the Eisteddfod certificate for Urdd Gobaith Cymru, one thousand copies of the *Blue Pilgrims* pamphlet, two sets of services, three sets of Christmas cards, a personal edition of engravings for Gertrude Hermes and the two books, *The Lovers' Song-Book* and *Theme With Variations*. In addition, Hodgson had printed all the wood engraved initials used to date as a specimen book for future use. A letter had been received from the author, Helen Waddell, giving permission for the Press to announce that her *Letters of Heloise* was in preparation.[1] Obviously sales of special bindings had increased as the Board decided to rescind their earlier resolution to limit these to fifteen copies per edition. In future, an ad hoc decision was to be made for each publication. The Secretary was instructed to see that type used for T. J.'s book and for the two abandoned works, *The Natural History of Selborne* and *The Book of Job*, was distributed on her return.

By 21 February, Dora Herbert Jones was again writing to T. J., this time to tell him that she had now decided to stay at the Press.

'I feel deeply bound up in this place, as you know, in the Press, in G.E.D. herself and, most of all, in the ideal Gregynog for which I have always hoped and worked, and like many another marriage, it has to be for better or worse, for richer, for poorer. I don't give lightly, or often, and you know I have given all I have to this place in various ways these last seven years.'

This period was to mark the beginning of an increasingly unhappy relationship between Gwen Davies and Dora Herbert Jones which lasted until Gwen's death. Despite a genuine affection on both sides, each came to mistrust the other. This was due entirely to a sad failure in communication. Clearly fault lay on both sides. Dora, hyper-sensitive concerning her apparent inferior position, was not a person with whom it was always easy to deal, and Gwen was basically too timid ever to deal with any difficult situation involving people satisfactorily, a fact T. J. realised all too well. He remained a stalwart friend and counsellor to both sides. Though deeply and sensitively sympathetic always, he was not above delivering a word of chastisement to either of these two women, who were his close friends, when he felt that it was merited. Indeed, his unfailing tact and diplomacy, borne of a deep insight into human behaviour, saved the Davies's from many an embarrassing situation in matters of personnel management.

Haberly's first visit to Gregynog as controller took place in late February. Gwen Davies afterwards reported to T. J. that she and Haberly had had a long *tête-a-tête*, as a result of which she felt he was going to be a great asset as he seemed to be more in sympathy with the aims of Gregynog than either of his predecessors had been. In early March, Haberly wrote to her, enclosing a copy of a page from Dante's *Divina Commedia* printed at Foligno in Italy in 1472 by Johann Neumeister of Mainz. He suggested that the type would be very suitable for the projected edition of Bridges's *Eros and Psyche*[2] and would harmonise well with the Burne-Jones illustrations which he intended to use. His letter ended,

> 'My life work is before me and I must close and be at it. What an opportunity you and your sister have given me: I can only say thank you for employing it.'

The matter of *Eros and Psyche* was discussed at a mid-March meeting of the Press Board, and Haberly's suggestion concerning the type was approved. As it would have to be redrawn for cutting, Haberly was requested to obtain estimates. The illustrations chosen by Haberly were to be culled from a set of about seventy drawings, prepared by Sir Edward Burne-Jones for William Morris's narrative poem, *Cupid and Psyche*. The majority of the blocks had been engraved by Morris himself, but the work was never published, and only a small number of the engravings had been published in other works. Tracings by various hands from Burne-Jones's original drawings were available in the Ashmolean Museum, Oxford, and Haberly announced his intention of redrawing a selection of these for subsequent engraving.

At the same meeting, the Secretary was instructed to write to Glyn Davies, Professor of Celtic Studies at Liverpool University, to enquire whether he would be willing to allow the Press to publish his translation of *Rugen Sange*, a collection of poems by the Danish poet, Jeppe Aakjaer. The request was later acceded to but with unfortunate results.

In April, Gwen Davies was somewhat taken aback when she received a letter from Christopher Sandford, now a partner in the Golden Cockerel Press, asking whether she would be prepared to sell the Gregynog Press as a going concern. It is not known what prompted this request, and, unfortunately, no copy of Gwen Davies's reply exists.

By May, the effect of employing a part-time controller was being felt in the Press, and Burdon Evans was growing increasingly concerned about lack of work to keep thirteen hands usefully employed. Haberly had not put in an appearance for some time. The truth was that he was hard at work on the illustrations for *Eros and Psyche* and did not want to be interrupted by having to visit Gregynog. He reported by letter that he had so far redrawn four of the Burne-Jones's illustrations and had sought the advice and approval of Robert Catterson-Smith who had been responsible for translating into line other illustrations by the same artist for the Kelmscott Press. As the redrawing was proving a much bigger task than he had anticipated, Haberly stated that this could not be undertaken as part of his duties as controller and requested an additional fee.

Haberly's attitude to his duties as Controller is demonstrated in a letter which he sent to T. J. on 6 May:

> 'It is now clear that I have as much as I can do in selecting and designing and overseeing without meddling in the mechanical work of the Press, which, so far as I have seen it, is excellent. If daily reports and proofs are sent to me, I can exercise all the necessary oversight without delay.'

The Press Board met again on 12 May. The Secretary was requested to write to John Beedham to enquire his fees for engraving, as it was thought he might be asked to work on blocks for *Eros and Psyche*. It was suggested that Haberly, who was present at the meeting, should arrange to travel to Denbigh to talk to Glyn Davies about the publication of his volume of poems before he returned to Buckinghamshire, a visit which obviously took place. The Press Board again discussed the appointment of one or more literary advisers. It was now many years since Ernest Rhys had acted in this capacity. Names put forward were Lascelles Abercrombie, Edmund Blunden and Edward Garnett. Haberly suggested Dr Joan Evans who was a personal friend of his. In the end, no appointment was made, and suggestions of titles for possible publication continued to come forward on a very arbitrary basis.[3]

One small task, which kept one or two of the staff occupied for some time in May, was the printing on vellum of a special copy of the Welsh League of Nations Union's *World Wireless Message of the Youth of Wales*. This was encased in a morocco roll, made and tooled by George Fisher, for presentation to the Prince of Wales on his visit to Llangollen on Goodwill Day, 18 May. The next two books had been completed, but were not yet published, and as June approached, all work was concentrated on preparing material relevant to the forthcoming second festival of music and poetry which was to be held 15–18 June. At the last moment, some of the soloists were unable to attend, so that alterations had to be made to the programme. Instructions concerning the changes were sent by telegram from Gwen Davies who was in London. Unfortunately, the telegram went astray, and by the time she returned to Gregynog, the printing was complete. Half of the thirty-six page programme had to be reprinted.

As though this were not enough, only a matter of days before the festival, it was discovered that moths had wrought havoc among the organ felts. An expert was flown over from Paris, and the felts were renewed in time for the event.

The festival followed the general pattern established in 1933: that is to say that it consisted of four main concerts and a religious service held on the Sunday morning. Sir Walford Davies and Adrian Boult were again the conductors, and, on this occasion, donations were requested on behalf of the unemployed.

A small string orchestra was assembled for the opening concert, held on Friday evening, which was dedicated to the memory of two distinguished musicians who had visited Gregynog in the past, Sir Edward Elgar, who had died on 23 February, and Gustav Holst, who had died on 25 May. The concert began, appropriately enough, with Lascelles Abercrombie reading Robert Bridges's poem, *In Memoriam*, which begins,

> 'Rejoice, ye dead, where'er your spirits dwell,
> Rejoice that yet on earth your fame is bright,
> And that your names, remember'd day and night,
> Live on the lips of those who love you well . . .'

The lines died away as the orchestra began to play *Nimrod* from Elgar's *Enigma Variations*. The choir then sang Holst's *O, Spiritual Pilgrims* which had been written for it the previous year. The first half of the concert ended with scenes from Elgar's *The Apostles* in which the choir were accompanied by Elsie Suddaby, Tom Pickering, Muriel Brunskill and Keith Falkner. After the interval, there was music by Elgar, Holst, Vaughan Williams and Brahms, and Lascelles Abercrombie read his poem entitled *Mary and the Bramble.*

The Saturday evening concert was devoted mainly to songs sung by Elsie Suddaby, Muriel Brunskill, Dora Herbert Jones and Keith Falkner and poetry read by Helen Waddell, Lascelles Abercrombie and Loyd Haberly. Sunday's concert consisted entirely of vocal works interspersed with poems by George Herbert and Henry Vaughan, read by Lascelles Abercrombie. The final concert on Monday was devoted to the music of Mozart, Schubert, Vaughan Williams, Walford Davies and Delius. Walford Davies had recently been appointed Master of the King's Musick, and Lascelles Abercrombie had written an ode in honour of the occasion which he now read. Two editions of the ode were printed by the Press, one a four-page pamphlet and one a broadside on Japanese vellum. A special copy of the broadside was printed on vellum and encased in a morocco roll, made by George Fisher, which was presented to the composer, together with a silver tray. Adrian Boult, who had married in 1933, was presented with a belated wedding gift of silver goblets by members of the choir.

Alas, for poor Gwen Davies the second Festival had fallen a little flat. In addition to her initial disappointment over some of the principal performers being unable to attend, T. J. had confessed that he had found the event disappointing and had gone away disgruntled and disapproving, which had hurt her deeply. Moreover, her old friend Walford Davies, normally so encouraging, failed to send her any word of appreciation after his departure. She brooded on the matter for some days, but, happily, her disappointment did not deter her from planning future festivals.

By July, Loyd Haberly had completed negotiations with Graily Hewitt to design a set of initials for the volume of Glyn Davies's translations. These were drawn for reproduction by line block as Hewitt was averse to drawing originals to be engraved by another hand. He was also contracted to redraw and supply missing characters for the Neumeister type, to be used for *Eros and Psyche*, which was to be cut by the Monotype Corporation, under Hewitt's supervision. No less a personage than Edward Johnston had approved the use of the type, so the Press Board felt that it was proceeding along the right lines.

Meanwhile, Haberly had also engaged the artist, Dorothy Hawkesley, to help him with the redrawing of the Burne-Jones designs for the illustrations.

The redesigning of the Neumeister type was the subject of an extensive correspondence between Hewitt and Haberly. Some fifty-three letters in all are extant in which matters concerning points of stress, serifs and other aspects of design are discussed in considerable detail. The work moved swiftly in Hewitt's hands, and by 14 September, he had supplied a complete set of drawings for the lower case alphabet. At a Press Board meeting held four days later, the directors agreed to accept Monotype's quotation for the cutting of ninety punches and preparation of the matrices at a cost of one hundred pounds, but they stipulated that Monotype should be asked to requote for accents and additional characters required for printing in the Welsh language, before the contract was signed.

Don Quixote: an introductory essay in psychology, by Salvador de Madariaga, which is dedicated 'to Tom Jones', was published in mid-September. Apart from a sincere liking for Madariaga as a man and an admiration for his writings on the part of T. J. and the sisters, it is hard to imagine why they should choose such a textbook-like work for publication. It is not at all the kind of thing one would single out for fine printing. It had the advantage, at that particular time, however, of not requiring any illustrations. As a piece of printing, it is a pleasantly competent if unremarkable book, set in Baskerville type, with initials and running heads in blue, printed on Batchelor hammer and anvil paper. The binding is very uninspired, being in quarter linen, with printed spine label, and blue-grey paper sides. The design for the fifteen copies bound in light blue levant morocco was devised by George Fisher. The decoration, which stretches across both covers, is composed of interlacing lines of gilt and blind tooling.

The prospectus for *Don Quixote* announced the forthcoming publication of Glyn Davies's translation of Jeppe Aakjaer's poems, with the title *Songs of Rye and Heather*, and also of an edition of Xenophon's *Cyrupaedia*. Alas, *Songs of Rye and Heather* was to die an early death, but not before it had been printed, for altercation arose between the translator and the Press which was to cause acute embarrassment. The Press Board, early on, incurred Glyn Davies's displeasure in that he had requested that the book be set in a Danish type face, a request which they had rejected on the grounds that such expenditure was unwarranted in the current economic situation. A little later on, Davies stipulated that no capital letters after the first should be used in the setting of the poems, which meant that the decorated initials, designed for the book by Graily Hewitt, would not be used. Haberly and Hewitt were in favour of retaining capitals, so no heed was paid to the translator's requirements. Worse still, Haberly had effected some rearrangement of the order of the poems. Proofs were prepared and approved by the Press Board who backed Haberly's decisions. Not altogether surprisingly, Glyn Davies regarded the Board's action as an infringement of his rights and in September requested that his agreement with the Press be terminated, that the type should be distributed and all proofs returned to him. The Press were tardy in acceding to his request, with the result that he wrote to the secretary on 6 October, accusing the Press Board of all manner of discourtesy:

'I would not consider the preposterous pretences and make-believe of your Board for a minute. There is a touch of arrogance, petty meanness, small-mindedness and ill-breeding about its views that settled the question for me at once. I am not giving any press the right to play the fool at its own discretion with any work of mine, not even a university press, much less a board of unskilled outsiders. Your Board was stupid to imagine that I would submit to the humiliation of having to subordinate my judgment to the vulgar whims of a young novice playing with his first big toy, who did not know the language of half the poems he was supposed to put in typography appropriate to the matter, and whose performance on some of the English texts I shall be glad to show anyone who wants to see what can be done in flamboyant bad taste.

Your Board should have said either YES or NO to my terms and not discussed them. But apparently they could not resist the attempt to be foolish. I must put them in their places . . .

The Board will only consult me as a matter of courtesy. I have had some examples of the courtesy of the Gregynog Press in the sending out of circulars without telling me, in altering my title in spite of my protests, an impudent interference with text, in altering my proofs without one word of warning, in submitting my copy behind my back to some ass who improved the spelling in defiance of the plain instruction FOLLOW COPY, and in ignoring my repeated requests for proofs.

And when Mr. Haberly, a raw young gentleman, tells me that the Latin mode in verse printing, as chosen by Aakjaer, Professor Gwynn Jones, the late Robert Bridges and Constables, is affectation, preciousness and sham sophistication (whatever is meant by this jargon), am I to take it as a fair specimen of Gregynog courtesy? . . .

The Gregynog Press do not consider themselves as an ordinary printing house. Neither do I. But they are subject to the law of the land like humbler presses, and I stopped them from committing illegalities.[4]

The Board will be good enough to remember that I never asked them to print my book. It was they who asked me, but they did not disclose the fact that they intended to dictate terms. And what do they know about fine printing that they should express an opinion on the differences between Mr. Haberly and myself? And they actually demand that I should trust the Press after the display that came to my hands! I had some sympathy with the experiments of a novice at first, but that soon wore off . . .'

Pointing out that, a week previously, he had asked that all proofs should be returned to him and the type distributed of which, as yet, there had been no sign, Davies continued,

'Will you please do what I ask you forthwith? Please do not compel me to ask the Authors' Society to intervene. I am sorry that you personally should be the intermediary in this correspondence, but, as for your Board, they have not shown me ordinary courtesy, and I am not concerned about their feelings. Things like your Board make me ashamed of being a Welshman!'

The letter was, quite obviously from the ill-composed manner in which it was expressed, dashed off in a fit of white hot temper. One marvels that the document is not singed round the edges, but Davies meant what he said. Fault doubtless lay on both sides, but the incident does point to two inescapable facts: the Press Board was anything but an expert body and it was capable, on occasion, of an arrogant disregard for the feelings of authors, many of whom were so clearly flattered to have their work published by the Press that they often turned a blind eye to the attitudes of its board of directors.

The rupture was beyond repair, and acting on the Chairman's instructions, the secretary replied to Davies's letter, enclosing his original copy and stating that all proofs had now been destroyed and that the type would be distributed in the very near future. On T. J.'s advice, Haberly was not shown Glyn Davies's actual letter. No doubt he felt that this would have caused a rift between controller and Press Board.

Haberly had not actually visited the Press since May, a situation which had inevitably caused delays, irritation and misunderstandings. Once more, T. J.'s intervention was sought, as a result of which Haberly arrived at Gregynog on 21 November, in time to supervise the final stages of the production of *Anne Boleyn*. He was unwell and seemingly not in a very agreeable frame of mind, which caused a stir in the Press.

T. J. was on the receiving end of the controller's displeasure, for Haberly wrote to him accusing the staff of mismanagement and idleness and the secretary of incompetence. He claimed that he had found several proof pages of the ill-starred *Songs of Rye and Heather* which had been reported destroyed. For this he blamed Dora Herbert Jones, though she claimed not to know of their existence. It transpired that these were mere sample settings of odd verses, made when the paper was being chosen, which lay innocently in a cupboard!

Loyd Haberly's own book, *Anne Boleyn and Other Poems* was published at the end of November in the hope of catching the Christmas market. It was not a resounding success. Set in Bembo type and printed on Kelmscott hand-made paper, with decorative initials designed by Graily Hewitt, the flow of the pages is interrupted by an indiscriminate use of bright red and bright green, a colour combination much favoured by Haberly. The title-page has a pleasing setting, except for Haberly's Press device which looks unsuitably contrived. The book has a precious air about it and an amateurish one at that. It was not sent out for review, at the author's request:

'I have always considered my verse to be my own private concern and naturally wish no more publicity than is necessary.'[5]

One wonders, in the circumstances, why he had been so keen for the Press to publish his verse.

The ordinary binding is in full brown oasis, blocked in gilt on the front cover with the crowned falcon badge. The special binding, designed by Loyd Haberly, is also covered in brown oasis, but in a darker tone. The centre of each cover is blocked in gilt with the crowned falcon and surrounded by an outer frame of double gilt rules, with conventional corner ornaments. It is reminiscent of a school prize binding of earlier vintage.[6] Haberly insisted that the head of both bindings should be gilded.

Two books had been printed and three published during Haberly's first year of office, but there was still some anxiety about the state of the fine book trade. J. G. Wilson, Manager of Bumpus's bookshop, had felt obliged to sell off remaining stocks of *Psalmau Dafydd* and *The Stealing of the Mare* at three guineas and two guineas respectively, which represented a loss of two guineas on the retail price of each. On a more encouraging note, a full set of Gregynog titles had been shown in the Royal Academy's Summer Exhibition of British Art. The suggestion to include them had come from John Johnson of Oxford University Press.

In addition to the guests attending the festival of music and poetry, two names stand out in the 1934 entries in the visitors' book. William Morris's daughter, May, had come over from Llanidloes, where she was camping with a friend, to have lunch with the Davies's on 17 August, and on 26 September, J. L. Garvin[7] called to see the press of which he had so frequently and sympathetically written.

Loyd Haberly was summoned back to Gregynog for a meeting of the Press Board on 4 January when he reported that the cutting of the new type was complete except for the accented letters. During the preparatory stages, the type had been named *Gwendoline*, but a decision was now taken to change the nomenclature to *Gregynog*. It was disclosed that Graily Hewitt had been asked to collaborate with Haberly in designing a special music type which would harmonise with the new type face. The drawings for this were already complete, and the type was later to be cut by the Monotype Corporation. It was reported that review copies of books sent to the United States had been returned unopened, on account of the high import duty payable, and Haberly was requested to encourage friends across the Atlantic to lend their own personal copies for review. The Board discussed plans for publishing an edition of Sir John Fortescue's *The Story of a Red-deer*, and it was resolved to ask Dorothy Burroughes[8] to illustrate the book, a request to which she agreed in late February.

A second Board meeting took place in London on 7 March, at which it was announced that seven of the Gregynog special bindings were to be included in an exhibition of thirty fine British bindings to be held during the summer at Columbia University. The titles chosen were: T. Gwynn Jones's *Detholiad o Ganiadau*, *The Autobiography of Edward, Lord Herbert of Cherbury*, W. J. Gruffydd's *Caniadau*, *The Singing Caravan*, *Erewhon*, *The Revelation of Saint John the Divine* and *The Lovers' Song-Book*. It is not recorded by whom the selection was made. The minutes also record the appointment for a probationary period of John Ewart Bowen, as binder's assistant. Bowen had been for some time a pupil teacher in a local school but was anxious to gain some craft experience.

Throughout the months of the spring and early summer, there was continuing correspondence between Loyd Haberly and Dorothy Burroughes concerning the illustrations for *The Story of a Red-deer*, and expenses were furnished for the artist to visit Exmoor to study red deer at first hand. The illustrations were to be printed in up to four colours, but Miss Burroughes was horrified at the controller's request that the original art work should be drawn to actual size, a practice not generally followed and not previously used by her, although she was an experienced illustrator. Eventually, however, she was forced to accede to his requirements, albeit reluctantly. Drawings were submitted at frequent intervals for Haberly's comment and criticism, and proofs were later hand-coloured by the artist as a guide to the printer, but the book was not to be published until 1936.

In April, Herbert Hodgson was again engaged in printing special items on vellum. Two copies of the Welsh League of Nations Union's *World Wireless Message of the Youth of Wales* were produced for presentation to the two young royal princesses in May. At Walford Davies's request, two copies of his *Jubilee Anthem* were printed for presentation to the King and Queen at its performance in St Paul's Cathedral on 5 May, and George Fisher was recalled from his holiday in Hampshire to bind them.

The special bindings were falling further and further behind publication. Haberly's suggestion that the number of skilled staff in the bindery should be increased and that of unskilled operatives in the Press be reduced was ignored, but Fisher was sent to his old employers, Rivière's, to enquire whether they would be willing to supply temporary help in finishing the special bindings. Though Rivière's were short of work at the time, none of the staff was apparently willing to help out at Gregynog. C. H. St John Hornby, manager at W. H. Smith & Sons' printing department, and Douglas Cockerell were both approached in the hope of acquiring temporary staff, but no help was forthcoming from either source.

At the suggestion of Henry Thomas of the British Museum, the Press undertook to publish his translation of *The Star of Seville*, a three act verse drama attributed to Lope de Vega. It was another curious choice of text, not exactly designed, one would have thought, to stimulate a dwindling market, for the style and tone of this early seventeenth century Spanish play of high-minded passion and death is unbearably stilted. It was published in May 1935, in time for Thomas to take two copies to Seville for presentation to the President of the republic and the Mayor of Seville on the occasion of the Lope de Vega tercentenary celebrations. The book was dedicated, at Loyd Haberly's suggestion, 'To Spain and to her greatest dramatist on the tercentenary of his death, this tribute from Wales.' It has a quietly satisfying title-page on which appears the arms of Seville, printed from line block. The arms were drawn by W. Charles Thomas, the translator's brother, who had recently designed the royal Jubilee medal. The text is set in Bembo and printed on Batchelor's GG watermarked paper, but the pages are marred by an undue amount of red ink, used for hanging titles, chapter numbers, names of characters and stage directions, which lends a disagreeably spotty effect, a feature not improved by the addition of bright green initial letters. A large and somewhat crude device, designed by Haberly, appears at the end of the book. The black oasis binding, blocked in blind with a Moorish style strapwork design and gilt central star, was designed by W. Charles Thomas. The inspiration for the design came from an old Sevillian binding in his brother's collection. (Henry Thomas was an expert on old Spanish bookbindings, of which he had the best collection in Britain.) From this time on, the majority of the ordinary bindings at Gregynog were covered in full leather in order to provide the necessary practice in finishing techniques for John Bowen, Fisher's newly acquired assistant. Three years were to elapse before Blair Hughes-Stanton was invited to design the special binding which is in purple levant morocco. It is decorated with onlaid blue-white stars, with beams of gilt lines radiating across both covers and spine, and has a base strip of onlaid vermilion oasis. The turn-ins are richly decorated with more vermilion onlay and gilt and blind lines.

A general atmosphere of unease was abroad in the Press during the summer of 1935, and Dora Herbert Jones made another bid to leave Gregynog in early June by applying for the post of Secretary at Royal Holloway College, in which she was unsuccessful. A report on the financial state of the Press, which W. J. Burdon Evans submitted to the sisters in the same month, was also far from encouraging, and for the first time since its inception, fears began to be expressed concerning the Press's future.[9] Gwen Davies still hankered after a resident artist at Gregynog, but her financial adviser was bitterly opposed to the idea, as a letter which he addressed to T. J. on 7 June reveals:

> 'Frankly, I am a little bit afraid of artists and other temperamental people. They so often run away with other people's wives or play shove-penny with the boys in their employ or talk sentimental tosh instead of getting on with the job. Sorry, but I'm so painfully normal.'

Painfully philistine too!

The Press Board met on 10 June at Gregynog, but if these more disquieting aspects of life were discussed, no evidence survives in the minutes. It was reported that the Golden Cockerel Press had been anxious to purchase the blocks cut by Gertrude Hermes for Gilbert White's *Natural History of Selborne* but had baulked at the price quoted which was fifty pounds, calculated on the cost of cutting. The meeting thus agreed to lower the price by twenty pounds, but the offer must still have proved unacceptable, as the blocks were never sold. Loyd Haberly announced that the new music type was almost ready and went on to suggest that it might be used to print a book of Welsh folksongs. The minutes finally record that John Bowen was to be offered a formal apprenticeship in the bindery.

Normal Press activity was again interrupted in early June by the need to produce, at fairly short notice, a programme and order of service for the annual Festival which was held 28 June–1 July. This year, preparations proceeded more smoothly than last. Sir Walford Davies and Adrian Boult were again much in evidence, though there were a few changes in the list of artists. Donations were this year directed to the Robert Jones and Agnes Hunt Orthopaedic Hospital and the Derwen Cripples Training College, both near Oswestry, in the neighbouring county of Salop.

On this occasion, the concerts followed a chronological sequence, beginning with a concert of Tudor and Stuart music, held on Friday evening. As a prologue, Lascelles Abercrombie and Zoe Havercroft-Jones, a local reader, read Bridges's *Ode to Music*, and the programme was composed of choral works, instrumental pieces and songs sung by Elsie Suddaby, Mary Jarred and Keith Falkner.

Saturday's concert was devoted to the seventeenth and eighteenth centuries and followed a similar pattern to that of the previous evening. The performers were joined by the Rothschild Quartet who played music from Gluck's *Orfeo*.

Sunday was full of event, beginning with a service, followed at ten-thirty by Part 1 of Bach's *Saint Matthew Passion*. Part 2 was given at eight o'clock. The choir, conducted by Boult, were accompanied by the Rothschild Quartet, Charles Clements at the organ and Michael Mullinar at the piano. The soloists were Elsie Suddaby, Mary Jarred, Tom Pickering, Keith Falkner, W. R. Allen and Cecil Cochrane. In between the two parts, the Rothschild Quartet played quartets by Mozart and Ravel.

The decision to split the Saint Matthew Passion into two parts,

XII

'If thou should cross the stream, which may not be,
Thou coudst not climb upon the hanging rocks,
Nor ever, as the goddess bade thee, see
The pasture of the yellow-fleecèd flocks:
Or if thou coud, their herded horns would gore
And slay thee on the crags, or thrust thee o'er
Ere thou coudst rob them of their golden locks.

performed at several hours' interval, was an unfortunate one, as Dora Herbert Jones later recalled:

'The singing that morning was superb: I think we were truly inspired. After the morning performance, an American visitor came up to me and said, "Anything could have happened this morning – *anything*", and you felt she was right. Part two, sung in the evening was as flat as anything. We should not have stopped but gone straight on.'[10]

The nineteenth and twentieth centuries featured in Monday's concert which opened with Walford Davies playing Mendelssohn's *Sonata for Organ. No. 1*, a particularly appropriate choice for the occasion as, among the guests, was Albrecht Mendelssohn-Bartholdy, the composer's grandson. Cecil Cochrane and Leila Megane, a young Welsh contralto, joined the other soloists for the remainder of the programme which consisted of pieces by Schubert, Schumann, Brahms, Bach, Holst and Walford Davies.

This year's Festival had gone much more smoothly than the last. No problems had arisen concerning artists, and it was generally agreed that the standard was high. To Gwen Davies's sorrow, T. J. was unable to attend, for his wife, René, lay seriously ill and died shortly afterwards. In writing later to console him in his loss, Gwen begged him to visit friends in America at her expense.

Loyd Haberly also absented himself from the festival. Already, it seemed, he was attempting to break his somewhat tenuous links with Gregynog. He later confessed to T. J. that he had been rather depressed by a visit from W. J. Burdon Evans who had not painted a very encouraging picture of the Press's future. His interest in running the Press does not appear ever to have been very strong, and this was the last straw. In September, he tendered his resignation which was to take effect from December. It was quite clear from his letters that he was much more interested in his other pursuits than he was in Gregynog. He was consulted as to his recommendations for the future development of the Press and suggested that the present staff, numbering thirteen, should be reduced, that a speciality should be made of music printing and that the Board should appoint a young man of education and taste as manager/controller.

Gwen Davies, meanwhile, had been shown some pages from the forthcoming *Eros and Psyche*, which was almost complete, and wrote enthusiastically to T. J. She was sure the book would cause something

Opposite: Page from Eros and Psyche, *with engraving by Loyd Haberly, after a design by Burne-Jones.* (*Reduced; side margins much reduced.*)

of a sensation. It did, but not of the kind she had envisaged! A month later, she was writing equally optimistically about the new music type, though she did feel that it was a little on the large side.

Eros and Psyche was published at the end of October. The text, printed in the new type in black and red, is further enhanced by green initial letters, specially designed for the book by Graily Hewitt. Haberly had wanted some of these to be in gold leaf, but Hewitt preferred colour. The type which has an open, rounded calligraphic quality is interesting but displays too many idiosyncracies to be admired. Lower case b and h are too easily confused; lower case w and x and upper case S and T are particularly bad designs, while the lower serifs in general leave much to be desired. Contemporary opinion varied: Bernard Newdigate thought it 'a magnificent type',[11] but C. H. St John Hornby, who was vehemently opposed to it, privately advised T. J. that it should be used no more. His opinion seems to have won the day, for apart from the 1935 Christmas card, where it was more successfully combined with the new music type in the setting of an old Welsh carol, it was not used again at Gregynog. Haberly, who was given a quantity of the type, later used it to greater effect for one book, *The Crowning Year*, printed in Dorset, and in books which he produced after his return to the United States.

It must be confessed that the type suffers badly in *Eros and Psyche* from the fact that it appears much too heavy in contrast to the lightness of the illustrations which, despite careful reworking by Dorothy Hawksley and the advice and help of Robert Catterson-Smith, appear sadly emasculated when compared with the blocks engraved on wood by William Morris after the Burne-Jones original drawings.[12] Haberly had selected twenty-four illustrations from the seventy or so tracings available, which were then photographed on to wood. He and Dorothy Hawksley had then inked over the lines, using a fine brush. Haberly himself engraved the title-page block and those on pages 3, 28, 45 and 112. The remainder were the work of R. John Beedham. The engravings appear to follow the Burne-Jones original drawings quite faithfully and, on that account, lack the strength and vigour imparted by Morris's graver. Whatever merit derives from the engravings, they fail dismally to match the weight of Hewitt's type. The size of the type was determined by the desire to make two stanzas of the verse equal in height to one of the blocks. Haberly had really intended the type to be used in conjunction with the special music type so as to render an effect similar to that of the early antiphonals.

The ordinary binding is infinitely preferable to the special binding. It is executed in white pigskin and decorated on the front cover by gilt blocked ornament within a circle and has raised spine bands tied down with gilt ties. John Bowen was responsible for the finishing which explains the hand tooling of the spine titling, as it provided the young apprentice with a good opportunity to practise and perfect his developing skill. Fisher designed the special binding for which a light green oasis was chosen to harmonise with the green initials in the text. 'The engravings and type,' Fisher later claimed, 'gave me a definite suggestion of silver. They also have a light dazzling quality, hence the decoration and top edge which is in a by-product of platinum.' The design, uninterestingly geometric, consists of three perpendicular bands of lace-like ornament which broadens to form a base, and similar ornament appears on the turn-ins.

The prospectus for *Eros and Psyche* included an announcement that an edition of *The Letters of Heloise*, translated from the Latin by Helen Waddell, was in preparation. Alas, due to the author's increasing domestic commitments, the work was not completed.

In early November, books from Gregynog were shown at an arts and crafts exhibition, held in Dorland House, Regent Street, London, where the Press shared a stand with the Halcyon Press. Robert Maynard was a member of the selection committee but asked permission to stand down while Gregynog items were being selected. Four items were eventually chosen – *The Singing Caravan* and *XXI Welsh Gypsy Folk-Tales* for their bindings, and *Psalmau Dafydd* and *The Fables of Esope* for their printing. These were impressively arranged against a black velvet background by Dora Herbert Jones who was in charge of the Gregynog part of the stand during the exhibition, and when Gwen Davies paid it a visit, she left well content. Seeing the Press books alongside ordinary trade exhibits strengthened her resolve to hang on to the Press at all costs. The fact that none of Haberly's books had been selected did, however, cause her to revise some of her views concerning the controller.

No sooner did she arrive back at Gregynog than she began to put some of her ideas for the future of the Press on paper to T. J. Chief among these was that the Board might consider appointing a panel of experts to advise on production in general. For matters relating to typography and layout she suggested C. H. St John Hornby, for the selection of artists Hugh Blaker, for literary appraisal of texts Lascelles

Abercrombie, Joan Evans,[13] Idris Bell[14] and others. It does not seem to have occurred to her that, not only would such machinery be unwieldy, but that few of the people mentioned would have either the time or the inclination for such consultative work.

There was also the problem of finding a replacement for Loyd Haberly yet to be solved. Names of likely candidates considered included many of those which had come up previously. To these was added the name of Adrian Bury, the painter, poet and writer on art, who was a friend of Hugh Blaker. All were found wanting in one respect or other, and none was deemed suitable for the job. Even as late as mid-December, there was still a possibility that Haberly's contract might be renewed. Haberly's offer to stay on for six months for a fee of £450, which he must have known the Press Board would not accept, seems to indicate his desire to be free of Gregynog. Even so, the chance to experiment with the new music type appealed strongly. He had already planned to produce a book of Welsh melodies and an edition of John Fletcher's *The Faithfull Shepherdess* in which he intended to use the music type in conjunction with Gregynog type, but neither of these publications was destined to see the light of day.

Two Christmas cards were printed in 1935. One consisted of the old Welsh carol, *Cwyd, a Gwel yr Adeilad*, for which both the Gregynog type and the music type were used.[15] The cover is plain except for the greeting, the first version of the cover having been rejected by Gwen Davies. In a letter to T. J., which she wrote on 19 December, she described this as having,

> 'a red dragon, curled up like a hedgehog, under a green hill with Noah's Ark trees on top of it and a star above that . . . all done in those crude red and green printing inks which he is so fond of.'

This so obviously refers to the device which appears at the end of *The Star of Seville*, published in May, that one marvels at its apparent novelty to her. For once, she had the courage to voice her objection directly to the controller.

The other card was something of a new venture. It consists of Helen Waddell's poem, *New York City*, accompanied by a startling wood engraving by the Polish artist, Stefan Mrozewski.[16] The colophon records the visit of Mrozewski to the Press on 5 December to supervise the printing. The invitation to attend had come from T. J., and Loyd Haberly, not having been consulted, was understandably a little put out. Gwen Davies herself succeeded in resolving matters amicably by sending Haberly a tactful letter, explaining that Mrozewski had been asked to supervise the printing of the wood engraving only, so that, at the last moment, the controller agreed to be present also. Three hundred and fifty copies were printed altogether, two hundred and fifty for use as Christmas cards, while the remainder were numbered and signed by artist and poet for circulation to friends and business associates.

By 16 December, no mutually agreeable decision had been arrived at concerning the renewal of Haberly's contract, and he left the Press. Two further books devised by him were already printed but not yet published.

Cyrupaedia: the Institution and Life of Cyrus, by Xenophon, prepared from the 1632 translation by Philemon Holland, with the text newly edited for the Press by F. W. Bateson, was published in September 1936. This handsome folio, printed in black, with marginal rubrics in red, on Batchelor's hand-made paper, is without doubt Haberly's finest Gregynog book. The setting in Poliphilus, with ample margins, is admirable, to which Haberly's hand-coloured floriated initials add a gracious touch. The only discordant note is struck by the title-page which, with its royal arms and royal dedication, is less successful in that it seems overcrowded with capitals. Professor H. W. Garrod, reviewing the book in *The Observer* for 13 December 1936, described it as 'a book of lordly sumptuosity', and the phrase is well merited. Nevertheless, Garrod went on to complain of inaccuracies in the Greek orthography and punctuation used in the notes:

'. . . which might have passed unnoticed but that some god, wishing to destroy Mr. Haberly, inspired him to print these marginalia in the brightest red that he could find.'

The ordinary binding was finished by John Bowen. It is in dark green oasis with central and corner onlays of lighter green and red which are blocked in gilt with foliate ornament of Persian character. By contrast, the special binding, though ingeniously contrived and faultlessly executed by Fisher, appears ugly. The design was conceived by Lady Cartwright of Aynho, an amateur bookbinder with whom Loyd Haberly was acquainted. According to the minutes of a Press Board meeting held in March 1935, the design earned her the fee of ten guineas. I have been unable to trace the original design, but it seems that it was amended and rendered suitable for the covers by George Fisher. The binding is in a heavily grained dark brown levant, with both covers bearing a central circular ornament made up of interlacing strapwork, onlaid in six colours and enhanced with gilt and blind tooling. Covers and spine also have onlaid panel frames outlined in gilt.

There could scarcely be a greater contrast between *Cyrupaedia* and Haberly's last book, J. W. Fortescue's *The Story of a Red-deer*, published in November 1936. It was a curious choice, for it is a story for children, written by the author for a young kinsman. The round blackness of Cloister type was an appropriate choice and harmonises well with Dorothy Burroughes's simple outline illustrations, sharply printed in a few clear, bright colours which were mixed specially by Herbert Hodgson to match the proofs coloured by the artist. This is the only occasion on which illustrations were actually printed in colour at Gregynog. Some adjustment of type sizes and weights and of the overall spacing would have improved the title-page which has a large decorative device in green, showing two stags in combat, against a wooded hill.

The book is bound in a pleasant red cloth which is blocked in gilt on the front cover with a stag. The special binding, in reddish-brown oasis, was finished by John Bowen and designed by George Fisher. The covers are richly tooled in gilt with a diagonal band, made up of two different fir tree tools. Above and below are gilt lines which are carried over the spine. The lighter lines, in the upper half, represent rain and the heavier lines, in the lower half, ploughed furrows. The inspiration for the design was a local scene encountered by Fisher. When out riding one day in the rain, he found himself confronted by a grove of conifers, alongside which ran a newly ploughed field. The binding has one other feature of interest in that the fourth spine band only is raised, a device not commonly used by bookbinders of the period but much used during the last twenty years.

Loyd Haberly's term of office as controller had raised a number of problems, largely due to the fact that he had chosen to disassociate himself from work in the Press for most of the time, a situation which was not popular either with the Board or with the staff. Though Gwen Davies had initially been impressed by Haberly's individual character and excited at the prospect of the Press at last having a type of its own, it is clear that criticism of his work which had latterly been filtering through to her from various sources led to a final disillusionment. *Eros and Psyche* was an interesting experiment which had misfired. For the remainder of Haberly's work at Gregynog perhaps only his production of the *Cyrupaedia* has any claim to lasting distinction.

Opposite: Page from Cyrupaedia, *with engraved initial (hand-coloured) by Loyd Haberly. (Reduced.)*

THE FIRST BOOKE, TREATING OF THE LIFE AND INSTITUTION OF CYRUS KING OF THE PERSIANS, THE FIRST OF THAT NAME.

THE PROŒME.

I ENTRED SOMETIMES INTO THIS serious cogitation: how many democraties(1) have beene subverted by those who would needs governe the common-wealth, otherwise than in a popular state. How many also, as well monarchies(2) as oligarchies(3) are come already to nought by conspiracies of the commons. How many persons likewise who attempted as absolute princes to reigne, were either very soone utterly overthrowne; or else if they ruled any while,held in great admiration,as passing wise men and fortunate. Me thought againe, I had observed thus much, that whereas in private families, some masters have many servants in household, others very few: yet even those few they could never have wholly at commaund. I considered, moreover, that as the feeders of neate and horses, so all others bearing the name of pastours(4), may by good right be reputed the commaunders of such beasts as they have under their hands. But withall, I perceived evidently, that all these heards are more willing to obey their pastours, than men their magistrates. For the said heards goe whithersoever their heardmen direct them; and as they gladly graze in those grounds(5) to which they are by them driven; so they forbeare such,as they debarre them fro. And verily,they suffer their said pastours to use at their owne pleasure,those profits that arise from them. Neither have we ever knowne any such heards to have risen rebelliously against their pastours, either by way of disobedience, or in denying them the use of their fruits. Nay rather, more curst and hurtfull are beasts to all others whomsoever, than to their rulers and those who receive commoditie by them: whereas contrariwise,men set themselves against none more, than such as they perceive, goe about to exercise dominion over them. When I had well pondered these poynts in my mind, thus I

1. Popular states or free states menaged by the people.

2. Absolute states ruled by one prince. Kingdomes.

3. Seignories under some few rulers.

4. Heardmen.

5. Or pasture.

After working in England for some time, Haberly returned to the United States where he later produced upwards of thirty books, a number of them printed in Gregynog type, or Paradiso type, as he has preferred to call it. These later works testify to Haberly's not inconsiderable talent as a book designer. Like his predecessors at Gregynog he had been required to fit into a mould predetermined by the nature of the Press and its owners and circumscribed by the dictates of a somewhat inflexible Press Board. Clearly the association could never have been wholly satisfactory, and one sees, in retrospect, that Haberly's pragmatic approach was totally unsuited to the task he had been set. His wayward whimsicality was not a thing to be grasped and pinioned: he required freedom to experiment and to grow. Reminiscing with the author in recent years, he recalled his brief association with the Press as an interesting and entertaining experience. Time had erased from memory all but the pleasant interludes.

For the first time since the birth of the Press, the year had been clouded for Gwen Davies by persistent feelings of pessimism concerning its future. The fine book trade was still in the doldrums, and the Press had been steadily losing money for some time past. The year drew to a close with no controller in view and no firm assurance of being able to continue. There had been personal anxieties too for the sisters. Gwen had been dogged by ill-health for the first few months of the year, due to a recurrence of the blood disease from which she was to suffer for the rest of her life, and the summer had been clouded by the death of T. J.'s wife.

Towards the end of the year, the sisters were invited to receive honorary doctorates of the University of Wales but refused on the grounds that they had done nothing to merit such honour. J. F. Rees, Principal of University College Aberystwyth, sought the intervention of T. J. who was unable to move them and was forced to reply in terms of defeat, noting that the refusal was 'characteristic of their chronic modesty'.[17]

Gwen Davies's confidence was clearly undermined by the general state of affairs, and her last letter of 1935 to T. J., written on 22 December, makes this painfully obvious.

'I have so longed to help other people, only to find that when they need me most – I just fail them.'

NOTES

1. T. J. had made the acquaintance of Helen Waddell early in 1933 and rapidly became a fervent admirer of the author and her work. She had been invited to the festival in July, and it was almost certainly on that occasion that she agreed to prepare the work for publication by the Press.
2. A work based on the story told by one of the characters in *Legend of the Golden Ass* by Apuleius.
3. No fewer than 127 titles suggested for publication and ultimately rejected by the Press Board are recorded among the Press papers.
4. A reference to Haberly's omission to obtain permission from Oxford University Press who had been given first option to print six of the poems included in the collection.
5. Letter from Loyd Haberly to T. J., dated 2 December 1934.
6. The finishing of this binding was carried out by John Ewart Bowen who joined the staff in February 1935.
7. James Louis Garvin, *d.* 1947. Editor of *The Observer*, 1908–1942, and editor of the 14th edition of *Encyclopedia Britannica.*
8. Dorothy Burroughes, *d.* 1963. Artist and illustrator of children's books.
9. The accounts show £672 worth of 'specials' and £3,561 worth of ordinary copies still unsold.
10. Conversation with the author in 1969.
11. Review by Bernard Newdigate in *The London Mercury*, Vol. XXXV, February 1936.
12. For comparison of these the reader is referred to R. Dunlap's *The Book That Never Was*. Oriole Editions, 1971.
13. Joan Evans was already working on her translation of the forthcoming *History of Saint Louis*.
14. [Harold] Idris Bell, Keeper of the Manuscripts and Egerton Librarian, the British Museum, 1929–1944. Knighted in 1946.
15. The only other use of the music type was in an undated printing of the first three verses of *Psalm 11*, *Pourquoy son bruit*? (Why do the people furiously rage?).
16. Mrozewski, who had previously worked in Amsterdam, had produced a series of wood engraved portraits of notable Englishmen. His portrait of G. K. Chesterton had won first prize in 1934 in the International Exhibition of Wood Engraving in Warsaw.
17. Letter from T. J. to J. F. Rees, dated 7 December 1935.

14
James Wardrop

THE OUTLOOK FOR THE PRESS AT THE BEGINNING of 1936 was dismal, and one Welsh newspaper even announced that it was about to close down! Then came Herbert Hodgson's resignation which was to take effect from the end of February. He wanted to take his wife and children, who were now growing up, back to London, but he was also growing tired of the erratic flow of work in the Press. For weeks on end he would have little or nothing to do, and then, suddenly, urgent jobs would be required which necessitated his working overtime. This was a sad blow for the Press, as it was unlikely that a printer of similar stature would be found to replace Hodgson.[1] It was decided to send Idris Jones and his brother, Idwal, to a large printing firm for three months' further training, and the Garden City Press, Letchworth, was eventually selected. Meanwhile, John Hugh Jones was sent to the Central School of Arts and Crafts in London for a short course in book design.

W. J. Burdon Evans, fearful that Dora Herbert Jones would be left in charge of the Press pending the appointment of a new controller, was instrumental in removing her from the Press altogether. At the beginning of April, he took over the secretarial duties himself until John Hugh Jones's return, and Dora's duties were thereafter confined to those of private secretary to the sisters. Heartbroken at being cut off from the affairs of her beloved Press, she had no choice but to accept. Thus ended ten years' loyal service.

The search for a new controller continued, and innumerable people were consulted. T. J. even travelled up to Glasgow School of Art to interview two members of the teaching staff, but nothing came of his visit. Finally, in March, Idris Bell of the British Museum suggested the name of James Wardrop, a Scot who was on the staff of the Victoria and Albert Museum. Victor Scholderer, the eminent bibliographer, had been one of those consulted, and he had stipulated that the new controller should be, first and foremost a typographer. Gwen Davies, however, still hankered to appoint an artist to the post. Wardrop was neither a typographer nor an artist. He was an authority on manuscript hands, a calligrapher and part-time lecturer in book production for the London County Council. When approached, he was not prepared to give up his post at the Museum but sought a year's secondment to Gregynog. This the Museum authorities were unwilling to grant as the Press was a private and not a public institution. By June, however, it was agreed that Wardrop should be appointed as part-time controller for a trial period of six months.

No books were printed during the first half of 1936, though Burdon Evans had reported to T. J. in January that the text of John Fletcher's *The Faithfull Shepherdess* was currently being set and that Agnes Miller Parker was to be asked to provide the illustrations. The book seems to have died an early death for there is no further mention of it. More ephemeral printing continued, and once again the time came round for the printing of the Festival literature.

In late June, shortly before the Festival was to be held, tremendous storms raged in Montgomeryshire. Heavy flooding of the river Severn in Newtown had caused the town to be without water. Gregynog had its own private reservoir, and the sisters arranged for daily supplies to be transported in milk churns to the Newtown hospital.

The Festival was held 3–5 July, and three concerts only were held. Walford Davies and Adrian Boult once more presided. Most of the artists taking part were the same as in 1935 except that Robert Speaight joined Lascelles Abercrombie in the poetry readings. The mixture was much as before, consisting of choral works, songs, instrumental pieces and poems.

For some reason the 1936 Festival came and went quietly. No mention of it whatsoever remains in the extant letters of either Gwen Davies or T. J. It was, however, reviewed by A. K. Holland in *The Liverpool Daily Post* of 8 August, and the Press reprinted his article shortly afterwards. This is how the Festival struck Holland:

> A visit to Gregynog is something in the nature of a pilgrimage. Here, amid the quiet hills of Montgomeryshire, in surroundings of singular charm and peacefulness, where nature and art vie with each other for the crown of repose, is enshrined annually a festival of poetry and music. The beautiful estate, the house with its treasures of classical and modern painting, everything in the atmosphere of the place contributes to that rare experience, art for love's sake.
>
> Here are none of your ordinary festival excitements. The programme, so carefully devised, a commemorative one in praise of famous men living and dead, and especially

those who have passed away during the last year, is not for sensation or that performance-interest which occupies so much of our time elsewhere. Here, music is made, poetry is spoken, with the agreeable intimacy that should belong to those arts. They become here familiar experiences, happy revelations, without ulterior motive, unless beauty and fitness are ulterior motives.

This year's festival . . . was the fourth of its kind. The programme, beautifully printed by the Gregynog Press, comprised three concerts, together with the Sunday service, which makes the whole festival seem like an act of devotion. But, as the choir sang at the opening concert, 'Nothing is here for tears . . . nothing but well and fair'. Brahms's 'Song of Destiny', Elgar's 'Music-Makers' were found in company with readings from Milton and Shirley, with Stanford's jovial 'Songs of the Fleet'. Rather than concerts in the ordinary sense, the proceedings could best be described as *consorts* of music and poetry, readings from the poets whom we have lost during the year past (Kipling, Chesterton, Housman, A. E., and others) interwoven with songs and choral pieces and music for strings, in delectable counterpoint. And that most delectable little opera of Vaughan Williams on Bunyan's text, with a new prologue from 'The Pilgrim's Progress', found a place beside the Bach motets and chorales, and Sir Walford Davies's deep-felt setting of the 'Ode on Time', which invokes the services of narrator, choir and soloist.

The conception of the Sunday service is a grace to the whole festival. It is exquisitely ordered; the readings, by two distinguished artists, of the prayers and passages from the scriptures; the choir, half-screened by the magnificent El Greco 'Via Crucis', which rivals the example in Toledo Cathedral; the wisely chosen music of Palestrina, Tallis, Brahms, and Mendelssohn; the hymns, and again the concluding 'Commemoration', a setting of 'Let us now praise famous men', and the poignant fragment, 'O Spiritual Pilgrim', which the late Gustav Holst wrote specially for Gregynog; the whole service breathes a unique spirit of poetry and mysticism. A hidden orchestra, conducted by Dr. Adrian Boult, who, with Sir Walford Davies, presides over the musical side of the festival, is playing the prelude to the second part of 'Gerontius', one of Elgar's most heavenly inspirations. It is an enchanted moment.

To speak of the many gracious and happy touches with which the festival is blessed, its intimacy, its charm of atmosphere, and the sense of community with which the distinguished artists (too numerous to mention) who take part in it manage to infuse their labour of love, would take much time and space. Suffice that Gregynog possesses a character all its own and provides an example that one could wish to see followed elsewhere.

During the summer, plans were formulated for an edition of the Arthurian legend, *Parsifal*, with a text condensed and translated from the German by Vera S. M. Fraser and wood engravings by Stefan Mrozewski. Towards the end of September, seventy-five engravings were printed by Idris Jones under Mrozewski's supervision. Acting on the advice of C. H. St John Hornby, it was decided early in 1937 that the text should be set in Matthias Claudius, an italic face designed by Berthold Wolpe[2] for the Bauer Foundry. The type was never acquired as the preparation of the text dragged on due to Vera Fraser's other literary commitments. In September 1938, she spent several weeks at work on it while lodged in one of the estate cottages in Tregynon and promised to have the text ready by the end of October. By December, the work was still not completed, but her nerves had been so shattered by the possibility of war that she asked to be relieved of her task. Yet another project, on which much time and money had been expended, was thus abandoned, and Mrozewski's engravings lie unpublished in the National Library of Wales.

Work in the Press in the latter half of the year was slack, and, in order to provide useful employment, a brief account of the Press, to which was appended a list of the thirty-four books published, was printed in the late autumn or early winter. (The item is undated.) The printing of Fulke Greville's *Caelica*, the first book to be prepared under Wardrop's direction, was completed in October but not published until 1937.

The last piece of printing for 1936 was, as usual, the Christmas card which consisted of G. K. Chesterton's poem, *The Wise Men*, enclosed in covers bearing a greetings message from the Gregynog Press, printed in black and red.

It had been another lean publishing year. No Press Board minutes are extant for 1936, and it would seem that such meetings as took place were rare and informal, though there is a reference in one of Gwen Davies's letters to the fact that Lascelles Abercrombie had been made a director.

The visitors' book for 1936 records the names, among others, of Vincent Massey, Canadian High Commissioner in Britain, and his wife who stayed at Gregynog 6–7 June, Joyce Grenfell who attended the festival in July, John Mason, first binder to the Press, who paid a visit on 3 September and Thomas Hennell, the artist and illustrator who was later to be killed in the war, who visited the Press on 22 December. Gwen Davies told T. J. in a September letter that hopes she had had of getting Bernard and Charlotte Shaw to Gregynog on their way from Malvern to Penrhyn had been frustrated and went on to relate an amusing anecdote of their progress across country. The Shaws had stopped for lunch at the Royal Oak Hotel in Welshpool, and when the proprietor had discovered their identity, he had been so overcome at playing host to so notable a pair that he immediately sent for the Mayor. However, the visit coincided with the Welshpool Agricultural Show whence the Mayor had taken himself. To his great chagrin, when he arrived at the Royal Oak, the Shaws had already departed.

One name disappeared forever from the list of visitors. Hugh Blaker, their friend and artistic adviser for many years, died of a heart attack on 28 September.

The Prime Minister, Stanley Baldwin, had been working at great pressure in the period prior to the abdication of King Edward VIII, and at T. J.'s suggestion, Gregynog Hall was lent to Baldwin and his wife from 7 August to 2 September as a quiet retreat, away from publicity. The sisters, meanwhile, stayed with their step-mother in Llandinam in order that their guests should be undisturbed. Before their arrival, Gwen Davies wrote to remind T. J. that Gregynog was a teetotal house where the only liquor available was a little brandy kept for medicinal purposes. She thoughtfully suggested that the Baldwins should be asked to place an order with Harrods and to charge it to her account. She afterwards reported to T. J. that they had ordered liberally – but on sale or return! The Baldwins' stay passed peacefully, attended only by the presence of the domestic staff, by Dora Herbert Jones, who carried out any necessary secretarial duties, and watched over by two detectives. Despite the Davies's vigilance, the visit had somehow been leaked to the press, so that the detectives were kept fully occupied diverting unwanted intruders. The sisters did travel over from Llandinam one afternoon to witness the planting of a tree by the Prime Minister at Gregynog as a memento of his visit and presented him with a copy of one of the Press books. The title is not recorded.

James Wardrop's contract was renewed in January 1937. A pattern was now established whereby the Press ran very much under its own steam, with secretarial duties and administrative matters being handled by John Hugh Jones and W. J. Burdon Evans and occasional visits from Wardrop at weekends and during holidays.

Fulke Greville's *Caelica*, edited by Una Ellis-Fermor, was published in April. One can only marvel that the Press Board should have chosen to reprint a collection of poems so unmemorable. The setting of the poems, in Perpetua type with red initials, is pleasant enough, but the title-page lacks unity. There is a lack of balance, and the letters are too widely spaced so that it has a pale and uninteresting aspect.

The ordinary binding which is very attractive was designed by Wardrop's wife, Evelyn. The broad dark green oasis spine piece has gilt titling and gilt lines extending on to the covers. The paper used for the boards, which is lighter in colour and patterned with black diagonal lines, was designed by James Wardrop and printed at the Press. The corners of the boards are leather tipped. Hughes-Stanton provided the design for the levant binding which is in slate blue levant morocco, decorated with a spangle of six-pointed tinted vellum stars, onlaid and outlined in gilt and blind, connected by diagonal lines of gilt and blind tooling.

In April, Gwen Davies learned that she was to be made a Companion of Honour. Her immediate response was to write to T. J., telling him that she wished to decline. His reply has not been preserved, but by 7 May she had obviously been prevailed upon to accept, since she wrote again saying that she was practising her curtsey every morning in the bathroom. The honour was duly conferred at Buckingham Palace, and she told T. J.,

'I felt I was representing the Gregynog team – just like the captain of a cricket XI.'[3]

Gwen had little heart for arranging the 1937 Festival of Music and Poetry, due to the fact that Dora Herbert Jones, the guiding spirit behind previous festivals, lay seriously ill in a Liverpool hospital where she remained from April until September. Thus the event was cut to two days and took place on 3 and 4 July. Perhaps the Press staff breathed a sigh of relief, as the programme is a mere eighteen pages in length. The reduced list of artists included the name of George Thalben-Ball, the celebrated organist.

After the national anthem, Saturday's concert began with the choir's rendering of Parry's *Ode At a Solemn Musick*, followed by an 'interlude of poetry and music' given by Lascelles Abercrombie and Walford Davies; Margaret Rees, another newcomer, singing songs by Purcell and Handel; and George Thalben-Ball playing organ solos by Handel, Bach and Michael Festing. The second half of the concert consisted of pieces for piano played by Walford Davies, songs by Rubinstein, Beethoven, Schubert and Purcell, sung by Mary Jarred, and *The Explorers* from Vaughan Williams's *Sea Symphony* sung by the choir, with Elsie Suddaby and Victor Harding.

Sunday evening's concert opened with the hymn, *O God Our Help in Ages Past*, and then the choir sang Vaughan Williams's *Te Deum Laudamus*, which was written for the coronation of King George VI. Organ pieces by Karg-Elert, Stanley, Yon and Boëllmann were played by George Thalben-Ball, and Lascelles Abercrombie read poems by himself, Henry Vaughan and Robert Bridges. The remainder of the programme consisted of music by Brahms, Bach, Handel and Düssek performed by the choir, accompanied by George Thalben-Ball at the organ, Elsie Suddaby, Mary Jarred and other soloists. The choir was, as always, conducted by Adrian Boult who had been knighted earlier in the year.

The History of Saint Louis, by John, Lord of Joinville, edited by Natalis de Wailly and translated from the French by Joan Evans, was published on 8 December. The book is one of the most impressive to come from Gregynog, and due credit must be given to James Wardrop for this sumptuous production to which several well known artists contributed. It is a folio, set in 16-point Poliphilus type with marginal rubrics in Blado, printed on hand-made paper specially made by Arnold and Foster.[4] The handsome initials and special openings, printed in red and light blue, were designed by Alfred Fairbank and cut on wood by R. John Beedham. The opening beginning IN THE NAME OF ALMIGHTY GOD on page seven is breathtaking. Seventeen large coats of arms were printed from wood engravings by Reynolds Stone and coloured by hand.[5] These appear at the foot of the page where they add a touch of brilliance. Two maps, at the end of the text, were designed and drawn by Berthold Wolpe. The map of France appears a little too finely drawn to harmonise well with the other pages; that of the Eastern Mediterranean is bolder and more in keeping. The title-page is a little disappointing. It is in monochrome and could have been improved in appearance by reducing in size the names of editor and translator and by moving them down slightly.[6]

One hundred and eighty-three copies were bound in dark maroon oasis, blocked on the front cover with the arms of Saint Louis in gilt. Fifteen copies, bound in mid-blue levant, were finished by John Bowen. The design is by James Wardrop and consists of a diaper of gilt fleur-de-lys on both covers and spine panels, outlined in gilt. The design is striking if somewhat stereotyped. Fisher commented,

> 'Wardrop was a traditionalist. It may be noted that the use of a section of pattern without adaptation to the confines of a book cover, as exemplified here, has failed to relate the spine to the sides and necessitated the cutting of eight extra tools.'[7]

By publication date, all fifteen copies were subscribed for.

Opposite: Page from The History of Saint Louis, *printed in red and black, with initials by Alfred Fairbank, engraved by* R. *John Beedham, and a coat of arms (hand-coloured) engraved by* R*eynolds Stone.* (R*educed.*)

IN THE NAME OF ALMIGHTY GOD

I, JOHN, LORD OF JOINVILLE, SENESCHAL OF CHAMPAGNE, HAVE CAUSED THE LIFE OF OUR SAINTLY KING LOUIS TO BE WRITTEN ACCORDING AS I SAW AND HEARD FOR THE SPACE OF THE SIX YEARS THAT I WAS IN HIS COMPANY ON PILGRIMAGE BEYOND THE SEA AND AFTER WE CAME BACK.

And ere I tell you of his great deeds and his knightly prowess will I tell you what I saw and heard of his godly sayings and his good precepts, so that they may be found set one after the other to edify those that shall hear them.

This holy man loved God with all his heart and followed His example: as appeareth in that even as God died for the love He bare His people, so put he his body in jeopardy many times for the love he bare his people; and he might well have avoided it an he had wished, as ye shall hear hereafter.

The great love that he had for his people appeareth in his saying to my lord Louis, his eldest son, in a dire illness that he had at Fontainebleau: 'Fair son,' said he, 'I beseech thee to make thyself beloved of the people of thy realm; for in sooth I had liefer have a Scot come from Scotland and govern the people of this realm faithfully and well, than that thou shouldst govern it manifestly ill.'

The saintly King loved truth so well that not even to the Saracens would he lie concerning his covenant with them, as ye shall hear hereafter.

So sober was he of his mouth that on no day of my life did I know him give thought to the ordering of any dish, as many rich men do; but he patiently ate whatsoever his cooks dressed and set before him. In his words was he temperate; for never in my life did I hear him speak ill of another, nor did I ever hear him name the Devil, which name is spilled about the kingdom, which methinks pleaseth God ill. He watered his wine in measure, according as he saw the wine would stand it. He asked me in Cyprus why I did not put water in my wine; and I told him that the physicians were the cause, who said that I had a thick head and a cold stomach, and that therefore I could not get drunk. And he

Bickers and Son of the Haymarket, London, wrote to the Press on behalf of a customer who wished to present a copy of the book to the library of the British Embassy in Paris. Gwen Davies agreed to the binding of a sixteenth copy but insisted that it should be different from the others in order to keep faith with the undertaking given in the prospectus. *Two* special copies were in fact bound. These are in cream levant with the arms of Saint Louis tooled in gilt, on a central blue onlay.

In December, the Press printed two hundred and fifty copies of Milton's ode and hymn, *On the Morning of Christ's Nativity*, set in Perpetua type, with a wood engraved frontispiece of the Virgin and Child by Alison McKenzie, a Scottish artist. It is one of the most attractive pieces of ephemeral printing ever produced at Gregynog and now commands high prices in the sale-rooms. Copies were numbered and distributed with an insert bearing Christmas greetings from the Davies's.

1937 had proved a fairly uneventful year at Gregynog but a quietly successful one as far as the Press was concerned. In thanking T. J. for the gift of a book for Christmas, Gwen Davies wrote on 27 December,

'I feel I can never thank you properly – all the accumulation of years; faith, courage, understanding and so much more.'

Conscious of the importance of the Press and confident of its standing in after years, the directors, from time to time, discussed the preparation of a history. T. J. was anxious that the collection of material for such a record should begin without delay and, early in 1938, urged that Dora Herbert Jones should be given leave to spend part of her time working on the project. W. J. Burdon Evans, no doubt a little jealous of her former position in the Press, was opposed to this idea, much to Gwen Davies's disgust. Writing to T. J. on 8 January, she commented,

'Of course, after all, Dora could only collect the material – *you* are the only one who knows it from the beginning. I wish W. J. wasn't so petty – he only likes people who are like machines and do as he tells them.'

One is left wondering why on earth, if this was really what she wanted, she could not have given the necessary permission, but she seems to have feared her financial adviser as she feared many another of her more important hirelings. The project was shelved, and T. J. planned to work on writing a full history of the Press after his retirement. Though, ultimately, he managed to produce no more than an outline history for his paper read to the Double Crown Club on 7 April 1954, it is to T. J. that posterity owes the preservation of many of the Press records which were carefully collected together before his death in 1955 and now form part of the private collection of his papers which is held on deposit in the National Library of Wales.

There was great excitement in the ranks of the Gregynog choir when on Tuesday 31 May, they travelled to London to take part in a Concert of the Four Nations, held at the Albert Hall, in which they were conducted by W. R. Allen.[8] A special programme, printed on Japanese vellum, was prepared by the Press for the occasion.

With Dora Herbert Jones restored to health once more, plans for the 1938 summer festival were entered into with much more enthusiasm than in the previous year, and the concerts were spread over the first three days of July. The list of artists taking part was much the same as before except that this year there was a new quartet, made up of Lady Schifner (oboe), W. H. Reed (violin), M. Gunyon (viola) and Eldrydd Dugdale (cello) who, in 1939, was to marry Lord Davies's eldest son and heir, Michael, later killed in the Second World War. Robert Speaight came straight from Canterbury where he was playing in T. S. Eliot's *Murder in the Cathedral.* Arwel Hughes, the young Welsh composer, also took part.

Friday's concert opened with a prologue, read by Lascelles Abercrombie, which was followed by George Thalben-Ball playing Handel's overture from *Alexander Severus*. Other items in the first half were the adagio and allegro from Mozart's *Quartet in F* for oboe and strings and the choir singing English part-songs and C. V. Stanford's *Heraclitus* and *Songs of The Fleet* in which they were joined by Keith Falkner. Part two of the concert consisted of Mary Jarred singing songs by Richard Strauss and Wagner and Vaughan Williams's *A Sea Symphony* sung by Elsie Suddaby and Victor Harding, accompanied by the choir, with Lascelles Abercrombie as narrator.

Saturday's concert began with Welsh poems and a group of Welsh airs arranged for the piano by Arwel Hughes. Music by Bach, Kodály, Mahler, Tschaikowsky, Walford Davies and Handel completed the programme.

The programme for Sunday requested that there be no applause due to the nature of the concert which consisted mainly of sacred works by Bach, Handel, Brahms, Elgar and Holst. During the concert, Robert

Speaight gave an electrifying reading of Becket's sermon from *Murder in the Cathedral.*

No one present could have realised that this would be the last festival of its kind at Gregynog, but by the summer of 1939, the general state of anxiety was such that all ideas of a summer festival were abandoned.

In addition to the festival programme, the Press also printed in June five hundred copies of *John Davies*, a memoir of the former Secretary of the South Wales District of the Workers' Educational Association who had died the previous December. Davies had been a frequent participant in conferences held at Gregynog. His dedication to the cause of education for those less fortunate than himself, his enthusiasm and unremitting hard work had long ago earned the respect of the Davies's who had counted him a friend. It was Gwen Davies herself who had promoted the project and who had commissioned Dora Herbert Jones to prepare the text. This thirty-six page pamphlet, printed in Baskerville type and bound in blue-grey paper covered boards, was issued for private circulation only.

Two books had also been in preparation during the summer, and both were published in November. The first of these was Henry Thomas's translation of the play by Eugenio Hartzenbusch, *The Lovers of Teruel.* It is a dismally dull play with a Romeo and Juliet type plot. Lascelles Abercrombie, when consulted as to the desirability of publishing the book, had replied,

> 'I should recommend the publication . . . provided the Press is willing to publish works which appeal not so much by their aesthetic quality as literature, as by their interest for students of literary history, a rather different matter.'[9]

The Press Board would have done well to heed his words. As a production, it is equally uninspired. The 16-point Bembo type appears too large on the page, though the effect is leavened a little by the initials, designed especially for the book by Alfred J. Fairbank.

The ordinary binding is in red niger morocco and decorated with panels of interlaced blind strapwork and has three false spine bands, outlined in blind. There are no records extant which mention the origin of the design, but its character suggests that it was probably the work of W. Charles Thomas. Blair Hughes-Stanton provided the design for the sage green levant cover of the first twenty copies. It is a striking design, executed in gold and blind tooling, with large geometric scarlet

Opposite: Page from The Praise and Happinesse of the Countrie-Life, *with engraving by Reynolds Stone.*

THE FIRST CHAPTER

Whoever loves the country, and lives in it upon his owne estate, whether hereditarie or purchased, and lends not his ears to any flat-t'ring allurements perswading to ambition and greatnesse, but carefully avoids those dangerous precipices and quicksands, I shall not feare to affirme, that such a liver is the wisest of men; for he, living upon his own, is no mans debtour, and is offensive to none but either a courtier or a citizen, and therefore is much more happy than if he had ingrost onlays, enhanced by a small black onlaid panel to which an intricate pattern of blind tooling forms a background.

The second work published in November was Antonio de Guevara's *The Praise and Happinesse of the Countrie-Life*, translated from the Spanish by Henry Vaughan and edited by Henry Thomas. This little book, the smallest ever produced at the Press, has a certain simple charm. It is set in Bell type, generously spaced, which harmonises well with Reynold Stone's delicate wood engraved head- and tail-pieces.

Engraving by Reynolds Stone, from
The Praise and Happinesse of the Countrie-Life.

The binding has olive green paper sides with a Reynolds Stone label, printed in black and red on pink, and a red niger spine, titled in gilt and decorated down front and back edge with a blind floral roll. The boards have minute leather tipped corners. It was the only Gregynog book to be issued in a dust wrapper. This bears a wood engraved title device by the artist. George Fisher devised the orange oasis binding of the first twenty copies in which the centre of both covers is blocked in blind with an intaglio block of eastern design, acquired in Damascus by Gwen Davies.

The final order of service to be printed at the Press in 1938 commemorated a sad occasion for the Davies sisters. This was a memorial service, held at Saint-Martin-in-the-Fields on 12 December, for their friend and adviser, Lascelles Abercrombie, who had died in October in his fifty-eighth year.

The 1938 Christmas card is a handsome affair, consisting of a single leaf, with the lines beginning 'Where shall wisdom be found?' from

Chapter 28 of the Book of Job, set in Perpetua type and beautifully printed on Batchelor hand-made paper which is loosely enclosed in grey paper covers.

By early 1939 the seeming inevitability of war was uppermost in everyone's mind. In April the sisters decided to move their valuables from their Buckingham Gate flat to Gregynog and suggested that T. J. should store his papers and other effects there too. The summer festival of music and poetry was abandoned, the Press Board ceased to meet, and books were talked about less and less.

A small handful of books was in preparation or in the initial planning stages. These included a translation of the poems of Dafydd ap Gwilym[10] by Idris Bell and his son, David, a work eventually interrupted by the war and never published. Ephemeral printing, while not at a standstill, was reduced to three orders of service, a programme for the choir's performance of Bach's *Saint Matthew Passion*, broadcast from Gregynog on 6 April, and a prospectus for the forthcoming *Shaw Gives Himself Away*.

Publishing a book by GBS was akin to walking on the rim of a volcano. As Gwen Davies wrote to T. J.,

> 'I feel we cannot be too careful about this book; it *must* be without a flaw . . .'

She went on to say that she was not at all sure that she liked the green tinted paper which Wardrop had chosen, even though the author approved. She reported that she had rejected the binding design which was a joint effort between James Wardrop and George Fisher. This she described as being,

> '. . . in bright sealing-wax red, with a lozenge-shaped medallion let in, with a little fancy tooling round it. Anything more banal would be difficult to imagine.'

The idea of commissioning Paul Nash to design the binding, 'even if he is dear', had occurred to her.[11] Nash visited Gregynog on 27 March, presumably to discuss plans for his designs. Ultimately, his design for the ordinary binding only was approved, that for the special edition being rejected.

A wood engraved frontispiece portrait, commissioned from John Farleigh, found little favour with the author who undertook to negotiate with the artist to improve the block. Shaw complained that it made his beard look black instead of red! Farleigh later claimed that alteration

Portrait of George Bernard Shaw engraved by John Farleigh, from Shaw Gives Himself Away.

to the beard alone accounted for five hours work. The block was finally amended to Shaw's satisfaction, though Gwen Davies remained disappointed with the portrait. Publication was delayed for several months due to the fact that Shaw insisted on the inclusion of one particular article which he had mislaid. The article finally came to light, and the book was published in November.

The book itself is unremarkable both from the production point of view and from the essays which it contains. These are not sparkling examples of Shaviana and are very dated. The book is perhaps best remembered for Nash's strikingly modern design on the ordinary binding. It is covered in a very dark green oasis, with an onlaid semi-abstract pattern based on the author's initials in orange, on the front cover, and, on the lower cover, a monogram of the same symbols in dark green on an orange coloured onlay. Both covers have a band of orange onlay across the foot. The construction is an unusual one to find in a Gregynog binding as it has a flexible back and no head- or tail-bands. The spine has one raised band only, above which is the title, imprint and date, blocked in orange foil.

Nash's rejected design for the special binding is an interesting one and in every way preferable to that chosen. It consists of a two-colour binding, the upper part in yellow and the lower part in purple, decorated with horizontal gilt lines. Both areas are relieved with onlaid discs of red and yellow and red and green which incorporate the author's initials.

There are two variant forms of the design eventually devised by George Fisher. Twenty-one copies are bound in the same dark green oasis, tooled in gilt with a design composed of perpendicular lines, varying in width, some of which are broken by a lozenge-shaped arrangement of graded lines which form a triangular pattern on each cover. Areas of similar pattern appear on the spine, and the whole is surrounded by a double rule gilt frame. The remaining four copies are in levant morocco and have a similar design, except that the frames are enriched with onlaid bands of light green and red.

A week after publication, Charlotte Shaw wrote to Gwen Davies to express their delight in the book, adding a typical Shaw comment,

> '[GBS] says there will be room for many scribbles and comments by him in his copy!'[12]

Gwen Davies's pleasure was to be short-lived, for in less than a month

Shaw himself wrote to say that he did not like the Nash binding. Her reply was a sad and stilted little note:

'I cannot tell you how much we regret your adverse criticism. We were particularly anxious that the volume should please you in all respects, and it is most disappointing to find that our purpose has not succeeded.'[13]

It had been hoped that the Press would have a collection of poems by Lascelles Abercrombie ready for publication before Christmas, but the type chosen by Wardrop was Jan van Krimpen's Romulus, available only direct from Holland. By early December, the type had reached the country but was still not at Gregynog, so plans were frustrated.

NOTES

1. Hodgson went to work for Robert Maynard at the Raven Press.
2. The type was later named Hyperion.
3. Letter from Gwen Davies to T. J., dated 10 May 1937.
4. The paper was made so as to print dry, but appears to have been damped.
5. Some of the shields are repeated, making twenty-five in all.
6. It ought to be pointed out that the initial planning for the book was carried out by Loyd Haberly before he left Gregynog. Joan Evans was a personal friend, and it was at his suggestion that the book was added to the Press publication list.
7. Conversation with the author in 1969.
8. W. R. (Willie) Allen was choirmaster to the Gregynog choir throughout its life and travelled over from Aberystwyth each Friday evening for rehearsals at the Hall.
9. Report by Lascelles Abercrombie, dated September 1937.
10. Dafydd ap Gwilym – Welsh poet, fl. 1340–1370.
11. Letter from Gwen Davies to T. J., dated 15 March 1939.
12. Letter from Charlotte Shaw to Gwen Davies, dated 13 November 1939.
13. Letter from Gwen Davies to George Bernard Shaw, dated 14 December 1939.

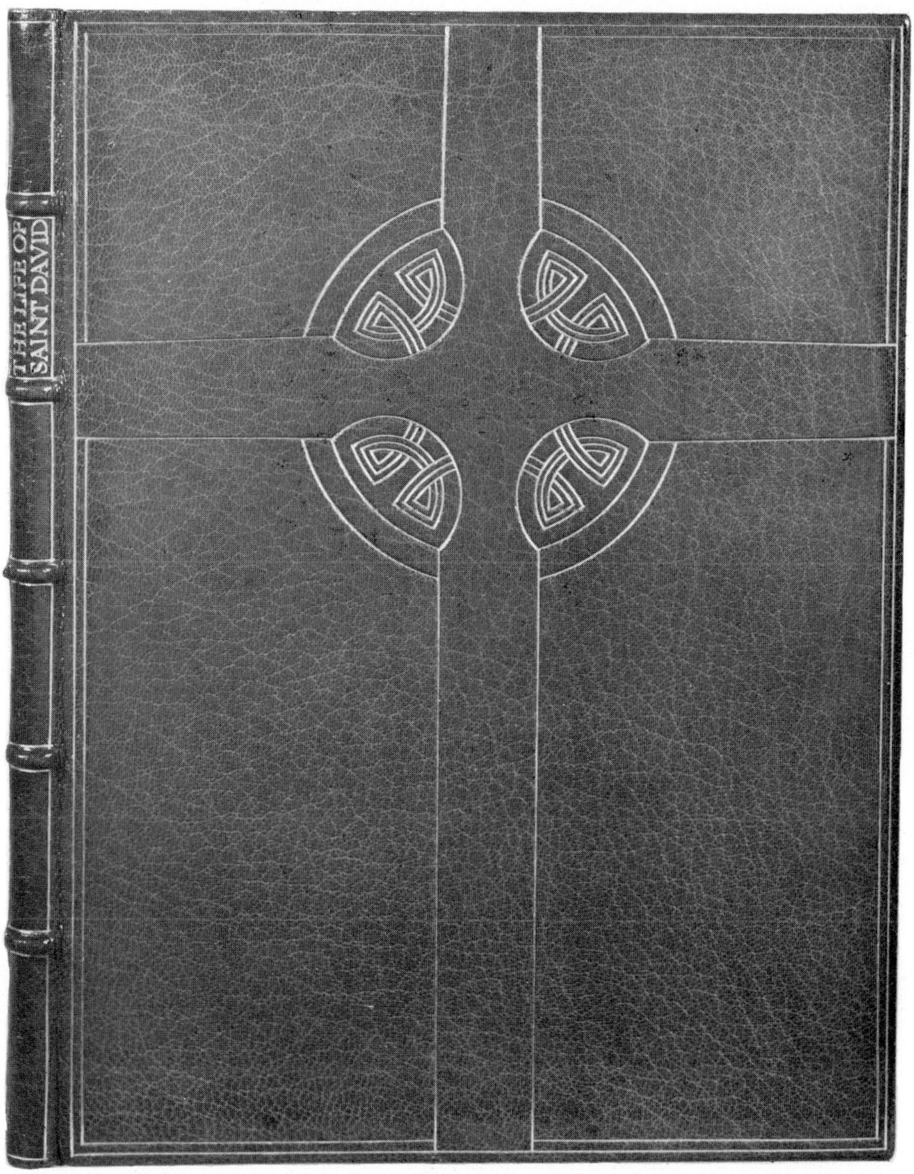

Special binding for The Life of Saint David, *1927.*
Designed by Horace Bray, bound by George Fisher.

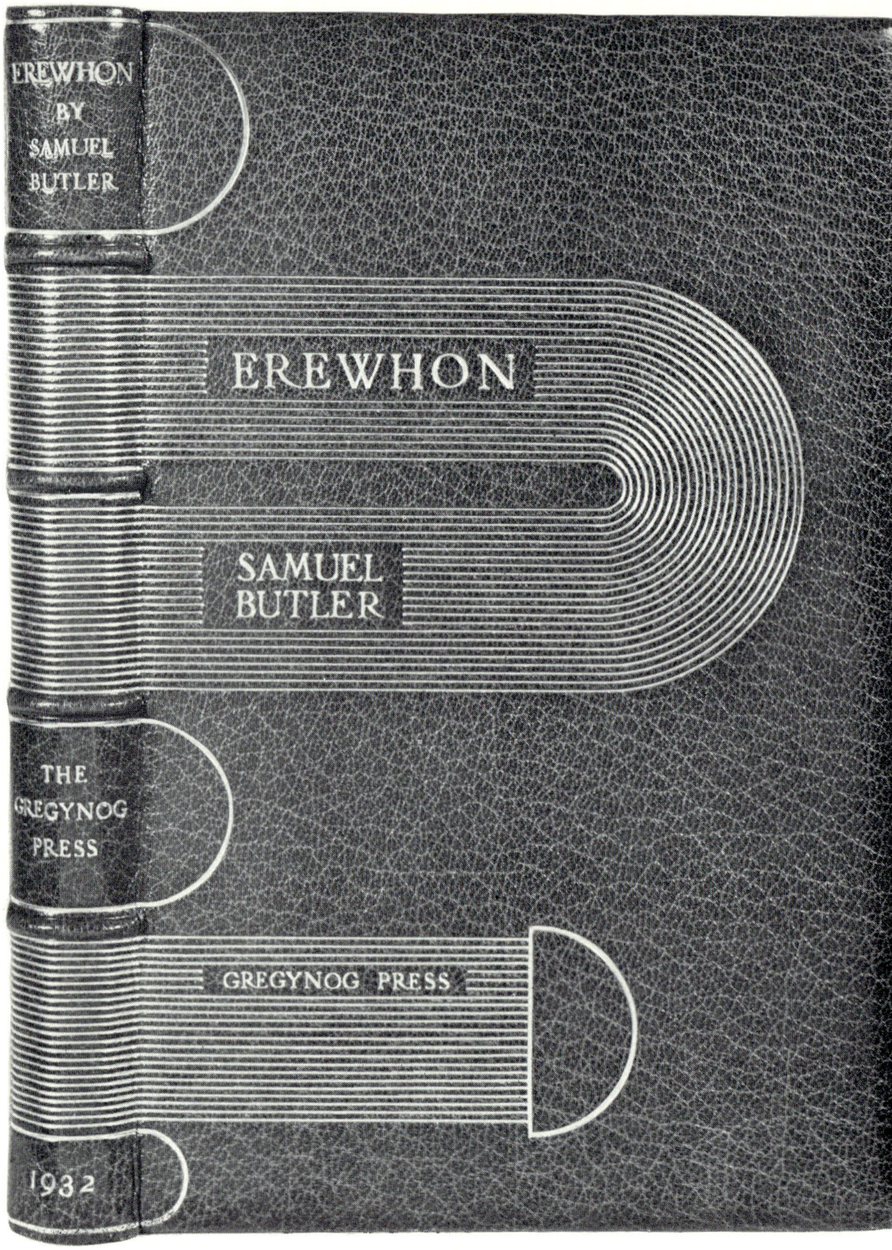

Special binding for Erewhon, *1933.*
Designed by Blair Hughes-Stanton; bound by George Fisher.

Opposite: Special binding for The Singing Caravan, *1932.*
Designed by William McCance; bound by George Fisher.

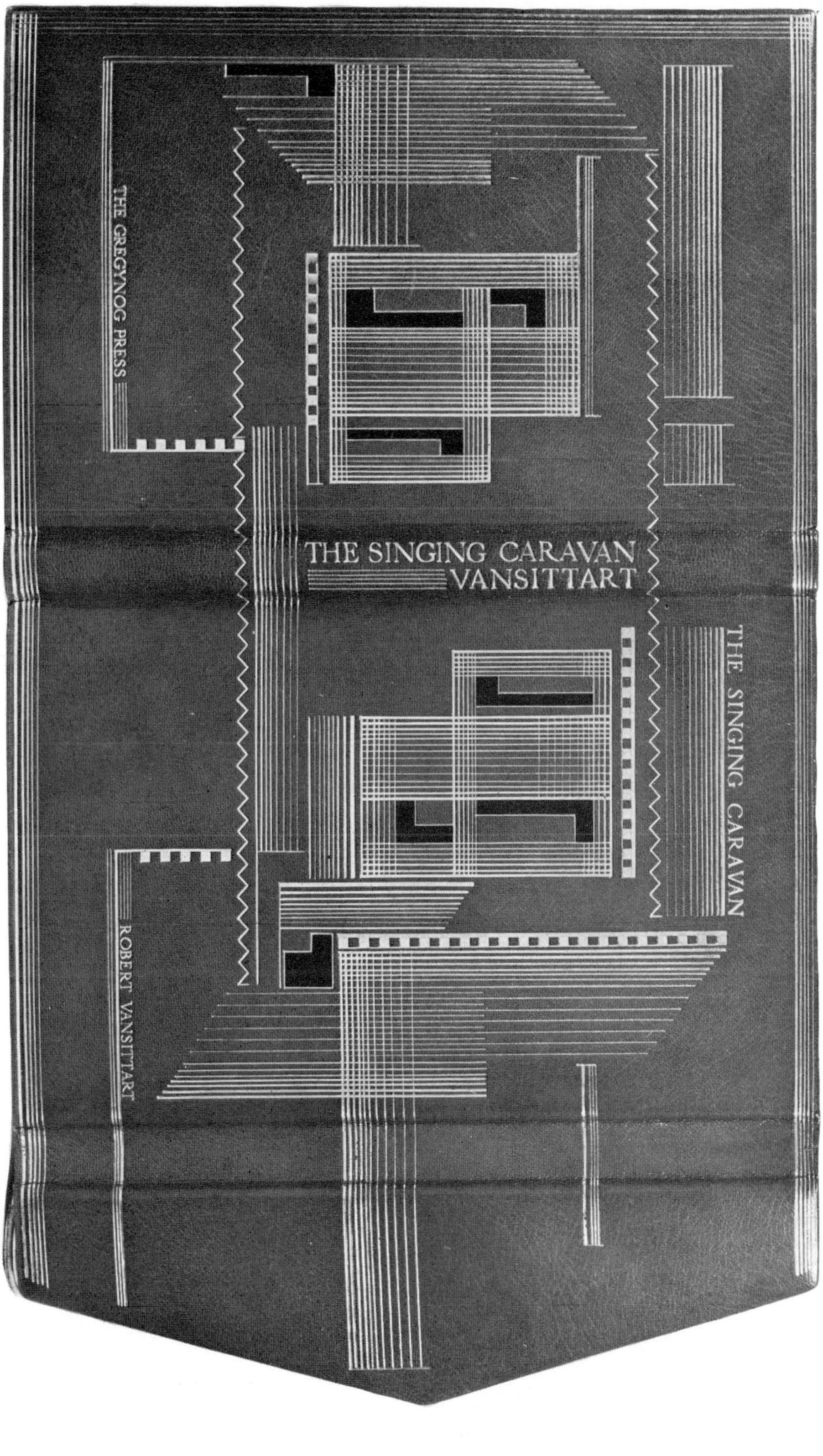
THE GREGYNOG PRESS
THE SINGING CARAVAN
VANSITTART
THE SINGING CARAVAN
ROBERT VANSITTART

Special binding for The Revelation of Saint John the Divine, *1933.*
Designed by Blair Hughes-Stanton; bound by George Fisher.

Special binding for The Lamentations of Jeremiah, *1934.*
Designed by Blair Hughes-Stanton; bound by George Fisher.

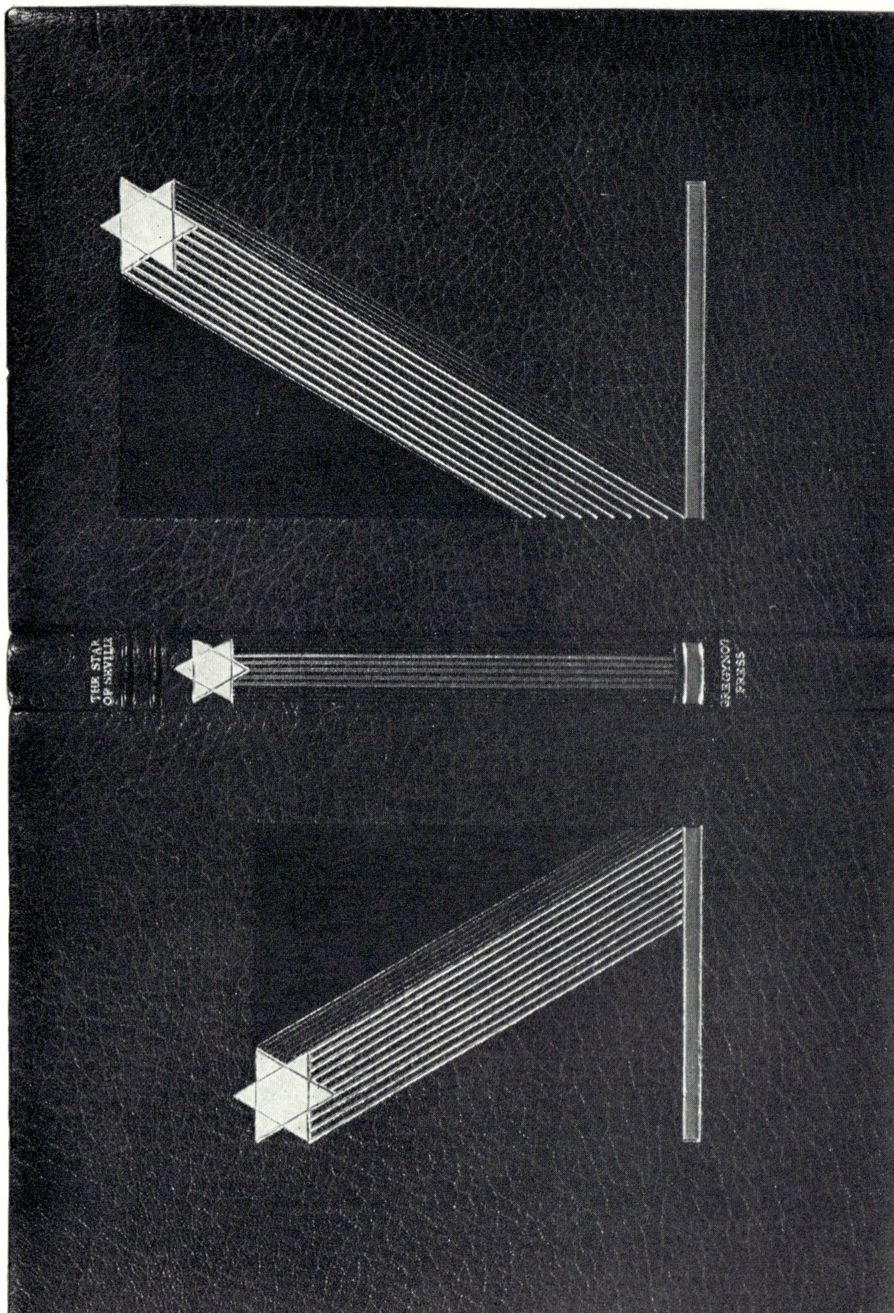

Special binding for The Star of Seville, *1935.*
Designed by Blair Hughes-Stanton; bound by George Fisher.

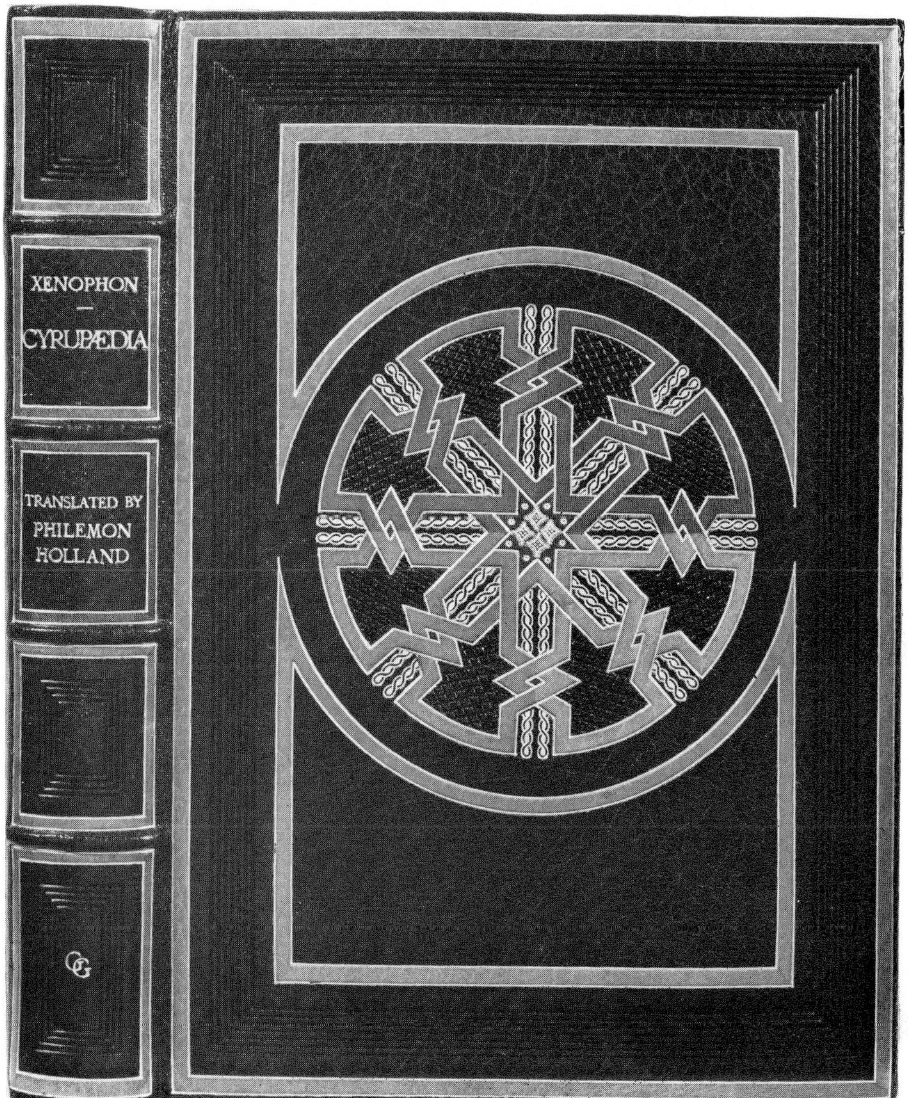

Special binding for Cyrupaedia, *1936*.
Designed by Lady Cartwright of Aynho and George Fisher; bound by George Fisher.

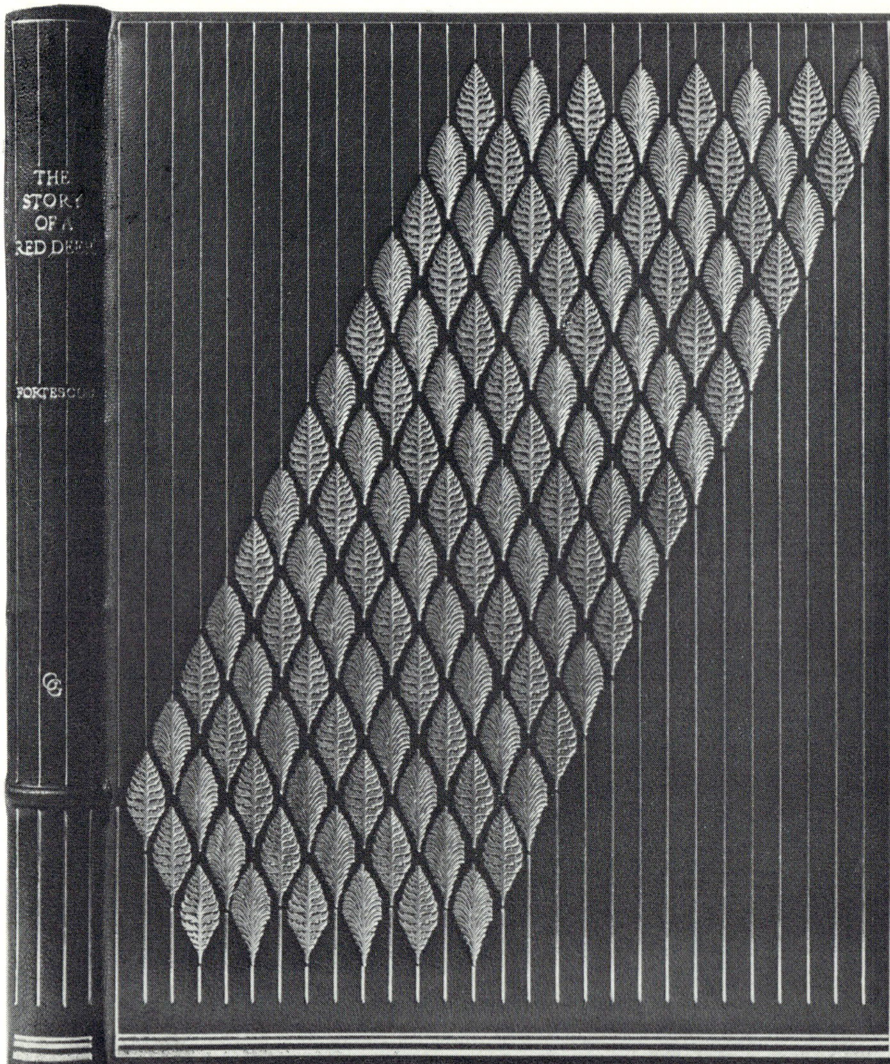

Special binding for The Story of a Red-deer, *1936.*
Designed by George Fisher; bound by John E. Bowen.

15
The End of the Press

THE FIRST SIGNS OF THE POSSIBLE DEMISE OF the Press came with the news in October 1939 that James Wardrop had been drafted into the Ministry of Supply where the strain of uncongenial employment caused him to be ill for several weeks. *The Praise and Happinesse of the Countrie-Life* had earned the Press adverse criticism from certain quarters. Burdon Evans was quick to seize on to this and on the fact that Wardrop, for some time past, had been unable to give the Press much of his attention. On 28 November, he wrote to T. J. pointing out that,

'The output of the Press for the past year has been deplorable, and the whole of the blame cannot be attributed to the world's prize egoist [GBS].'

In Burdon Evans's opinion, Wardrop ought to have offered to forego all or part of his salary and, in December, wrote to tell him so, blaming Wardrop for delays in the printing and publishing programme. Worse still, he withheld the controller's cheque which prompted poor Wardrop to write a letter in self-defence, offering to relinquish his services, an offer which Gwen Davies was loath to accept but felt she had no alternative.

Finally, Wardrop was asked to relinquish his post from the end of February, up to which time his salary would be paid. Hearing the news, T. J. wrote Gwen a sorrowful letter of acquiescence in which he commented bleakly,

'. . . but I do add that I regret that so rarely are we able to part with any Press employee in a quite unclouded atmosphere. Pity! . . . Meanwhile, let some of us love one another and forgive one another our trespasses, as we hope to be forgiven.'[1]

It is an awful indictment of relations which prevailed between management and staff at Gregynog that his words rang all too true, and it is a sobering fact that, ultimately, of all the principal staff, George Fisher alone left without rancour.

One cannot help but feel for Wardrop in restrospect when one reads a letter he wrote to T. J. asking to be allowed to see him:

'. . . as I am most anxious to avoid the imputation, above everything, that I have let down the Press in any way. I was conscious that my reputation had suffered a jolt after the publication of the Vaughan,[2] but I still think it is a good book, in which conviction I am fortified by the opinion of typographers, Stanley Morison not the least among them.'[3]

When Wardrop left the Press, two books were in preparation. Seven books in all stand to his credit, chief among them *The History of Saint Louis*, undeniably one of Gregynog's finest achievements. For the rest, they are competent productions, severely restrained, classical in style and otherwise unremarkable.

As early as 18 October, Gwen Davies had written to T. J.,

'I can't help feeling that this Abercrombie book will be the swan's song of the Press, and one would like it to be worthy of our *best* tradition.'

Yet a month later, replying to a letter in which T.J. had pointed out the impossibility of continuing without a controller, she wrote,

'I was bitterly disappointed that you should envisage the closing of the Press – almost as though it were inevitable – surely at a time like the present we should try desperately to keep the things that matter alive – the things of beauty, treasures of the mind and the joy of creating. It won't be possible to open the Press again ever once we close down. Forgive me, but oh, don't give up our dreams like this – without a struggle.'[4]

Now the prospect of closure hung like a menacing cloud, but Gwen Davies and Burdon Evans were determined to carry on for as long as possible and complete books in preparation, for, as Burdon Evans had rightly pointed out, once the staff were dispersed, it would be impossible ever to reassemble the old team. Idris Jones had already volunteered for service in the Royal Air Force, which left the Press without a printer. Robert Maynard was asked whether he would release Herbert Hodgson to print the last two books, a request to which he readily acceded.

It had indeed been a black year, in which one of the few bright spots had been an invitation from The British Council to exhibit three of the Press bindings in their Exhibition of Contemporary British Art in Stockholm, 26 April to 15 May. In May, T. J. reported in a letter to Dora Herbert Jones that he and Burdon Evans had attended a sale of books in Hodgson's Rooms at which Gregynog books were sold at depressingly low prices. Some of the Welsh books had realised less than one pound each. He and his companion had made bids for some of the Welsh items in order to push up the prices a little.

In July, T. J. had undergone a series of operations, performed by the royal surgeon, Lord Dawson of Penn, which had caused Gwen Davies much anxiety and distress. By August, however, he was on the road to recovery and spent a month's convalescence at Gregynog.

This year there was no festival to lift Gwen Davies's spirits. In December, Sir Adrian Boult made a broadcast from Bristol in a series entitled *Music In Wartime*, in which the Gregynog choir made its last major appearance. The programme included a talk by Boult on the history and activities of the choir.

For the first time since 1922 there was no Gregynog Christmas card.

To add to the general atmosphere of gloom prevailing that winter, heavy falls of snow and severe frosts in February wrought havoc among the trees in Gregynog park. Broken limbs and branches lay everywhere. Several of the Press staff had already been called up or had volunteered for the armed services; others were soon to follow. John Hugh Jones remained as, under the official regulations, in his capacity as compositor and printer's manager he was allowed to do. The much depleted staff, with Hodgson's help, completed the printing of Ellis Wynne's *Gweledigaetheu y Bardd Cwsc: Visions of the Sleeping Bard* in March and Lascelles Abercrombie's *Lyrics and Unfinished Poems* in June which were published in March and August respectively.

The first of these is a bilingual work, set in Bembo type. It serves to underline the problems inherent in bilingual printing, for the Welsh text occupies considerably less space than the English. Here a solution of sorts was found in using a larger type size for the Welsh, though the effect is not entirely satisfactory. The book has a handsome wood engraved frontispiece by Blair Hughes-Stanton which accords well in weight with the title-page but, unhappily, does not catch the spirit of the text. This book has one of the most pleasing ordinary bindings. It has a red oasis back, tooled down its length with seven gilt lines which separate title and imprint, and the sides are covered in a very attractive red cloth, powdered with gold. Fisher designed the special binding which is covered in large grain purple oasis, tooled in gilt with a series of upward pointing arrow-like forms, enclosed in a heavy gilt rule frame. Two only of the five bands are raised so as to form three gilt panelled spine compartments.

As previously mentioned, obtaining the type for the Lascelles Abercrombie volume had created problems. Even when the type eventually arrived it was insufficient, so that the type used for the first half of the book had to be distributed and used for the second half. The elegance of Romulus is well displayed on the smooth, green-toned Barcham Green paper, yet the page format is a little too wide for the narrow setting. The binding has green and black Cockerell marbled paper sides and a dark green oasis spine, down the three surfaces of which are gilt lines, terminating on front and back in a small lozenge and on the spine in the GG monogram. The special binding was again devised by George Fisher and is one of his more interesting excursions into design. The large grain blue-grey oasis cover is tooled all over with gilt lines of varying thickness which form an effective pattern.

This was, alas, to be the last Gregynog book. Ephemeral printing continued to emerge from time to time, but all the pieces lack the assured style and technique of earlier printing. The final appearance of the Gregynog choir was a choral recital given in the village hall at Llandinam on 6 December 1940 in aid of the Lord Mayor of London's Air Raid Disaster Fund. Many of the favourite items from earlier festival performances were repeated. They sang English part-songs of various periods, songs by Hubert Parry, C. V. Stanford and Bach, a group of carols and a chorus from Brahms's *Requiem*. Sir Walford Davies, their faithful supporter to the last, played piano pieces by Beethoven and Brahms. The programme is printed on a single folded sheet and bears no imprint.

Two other pieces of printing, each consisting of a single sheet, appeared at the end of 1940. At T. J.'s request, John Hugh Jones set up and printed seventeen *Resolutions When I Come to be Old* by Jonathan Swift, which was later distributed with the greeting, 'From T-J, Gregynog Press, New Year's Day, 1941'. The other is an acknowledgment from Dora Herbert Jones and family of condolences received on the death at sea of her daughter, Elsbeth. Elsbeth had sailed in the autumn as an escort on the Rangitane, a ship carrying evacuee children to Australia. On the return voyage, the Rangitane was sunk by a German raider. Elsbeth was the only casualty. She died of injuries and was buried at sea from the German ship on 30 November.[5]

There were no orders of service, no certificates and no Christmas card printed in 1940. Many of the special bindings, always some years behind publication, still lay in sheets. To keep faith with subscribers,

Fisher agreed to stay on at a reduced salary to finish them.

In May, about half of the Davies collection of paintings and their collection of early and fine books and manuscripts, together with a complete set of the Gregynog books, had been deposited at the National Library of Wales as an insurance against destruction by fire. To anyone who has visited Gregynog this would seem to have been an unnecessary precaution, but the house was scheduled to become a Red Cross convalescent hospital for members of the armed services which it remained for the duration of the war.

Sir Walford Davies, friend and supporter of the Davies sisters and joint inspirer with T. J. of the Gregynog venture, died on 11 March 1941. A memorial service was held in the Music Room on 19 April in which Sir Adrian Boult and Elsie Suddaby took an active part, and the Press printed the customary order of service. This is the last piece of printing to emanate from Gregynog before the Press was properly disbanded.

Now caster and press lay silent. Only John Hugh Jones remained in an administrative capacity. Above, in the bindery, George Fisher worked on until 1945, with the occasional help of Gwen Edwards, the sister of Idris and Idwal Jones, who had been one of his chief bindery assistants since 1927. In the artist's studio adjoining, Daisy Davies and Jane Blaker were often to be found putting strengthened covers on to paper bound books prior to their despatch to the armed services. Everything remained just as it had been, awaiting the day when peace would bring the Press to life again.

John Hugh Jones's continued presence in the Press irked W. J. Burdon Evans, and on 24 April 1941, he sent Jones a letter which would seem to have constituted an unwarranted attack on his personal integrity. The letter urged him to join up without delay and went on,

> 'You will appreciate that it is not merely a case of individual responsibility, but you are making us a party to your immobilisation when every man is wanted. Meantime, I feel that we ought to reduce your salary still further.'

Stung by this, Jones decided to leave Gregynog in mid-May, before any fresh attack could be launched from the same quarter. Thus, unhappily, ended more than eighteen years' loyal and devoted service to the Press, service which appears, in many ways, to have been held in small regard by those who should have known better. Burdon Evans's action was, nevertheless, deplored by Gwen Davies who wrote Jones an appreciative farewell letter. One more tribute to his services came from James Wardrop,

'I have always had a high personal regard for [Jones] and remember with gratitude his unswerving loyalty to me at the Press.'[6]

For some time, cash payments were made by the sisters to members of the Press staff serving in the war but not to John Hugh Jones who, after unsuccessful attempts to join the Royal Air Force, went to run a boys' club in Bristol.

Once more T. J. tried to persuade Gwen Davies to allow Dora Herbert Jones, who was still acting as secretary to the sisters and hospital secretary for the Red Cross, to prepare a draft history of the Press, but his request met with little enthusiasm. Referring to the proposed project, she wrote,

'It is all rather a melancholy subject – just like writing someone's obituary notice – but I fear the days of Gregynog's glory are at an end.'[7]

The history was never begun. Mutual feelings of affection, tinged with mistrust, had for many years characterised the relationship between Gwen Davies and her secretary. Now that they were forced by circumstance to be constantly in each other's company, proximity proved increasingly irksome. In March 1942, Dora accepted a post with the Ministry of Labour in Cardiff. It was with considerable misgivings, however, that she contemplated leaving Gregynog where the Press, the services, the music and the general ambience of a cultural milieu, not to mention her very real affection for Gwen Davies, had meant so much to her. Highly sensitive and oft-times temperamental, she had not always proved as malleable an employee as the Davies's would have liked, yet her contribution to the realisation of many of Gwen Davies's dreams for Gregynog had been invaluable. The atmosphere between the two women had grown so tense that Gwen believed that to accept Dora's resignation was the only way to relieve the situation, yet she grieved for all that she stood to lose by so doing. To T. J. she wrote,

'. . . it seems the end of so much of me . . . If ever there was a dual personality, it is Dora – on the one hand she is aggressive, self-assertive, opinionated, domineering, touchy and suspicious . . . while the other side, which she shows to *very* few, . . . is affectionate, generous, warm-hearted, able, interesting, full of imagination and idealism and enthusiasm, full of poetry and music and the lovely things of earth – in fact, the Dora of the services and the Festivals – the Dora with whom I planned and made Gregynog and all it has meant in the life of Wales. I shall now have no one to talk to or plan with. Daisy and I are so different – we live in different worlds almost – that is the tragedy of it.'[8]

It was indeed a tragedy: the loss was mutual, and in later years they were to meet but rarely, each mistakenly assuming that the other had turned against her.

Just before Dora left Gregynog, on Easter eve, Mrs Edward Davies, the sisters' step-mother died. During her last illness, the news of Dora's impending departure, of which she would never have approved, was carefully kept from her. Gwen's cup of unhappiness was full.

Despite the war, sales of the Press books continued fairly steady. By mid-August 1942, Colonel E. C. Powell, recently appointed as general assistant at Gregynog, reported that £410 worth of books, representing one hundred and fifty-eight volumes, had been sold since the beginning of April when a list of the twenty titles still available was sent out to subscribers and others. By January 1944, only seven titles were still available, and the following June, at a meeting of the Press Board, it was reported that only three hundred and fifty-eight books in all remained to be sold.

On the personal front, 1944 was a rather bleak year for the Davies's. Their brother, Lord Davies, died from cancer on 16 June. Gwen, with whom he had been frequently at variance over the years, wrote to T. J. the same day,

'God pardon his faults and remember his virtues.'

He was cremated in Birmingham on 20 June and his ashes scattered by the family the same evening, on a hillside above Llandinam. He was succeeded as second baron by his eldest son, Michael, who, surviving him by months only, died of wounds in Holland in September. Nor was all peaceful at Gregynog itself, for, in August, one of the patients had committed suicide by throwing himself from an upper window, and his body had landed outside Daisy's sitting room.

In October, T. J. was elected President of University College Aberystwyth, and from 1945 until 1954 made his home in the town. Physically nearer now to the two sisters, he continued to help and advise them in a variety of matters. They, meanwhile, contributed to his creature comforts by sending him consignments of logs.

By April 1945, all the special bindings were completed, with the exception of sixteen copies of *The Plays of Euripides* which were finished in 1952 by John Bowen at the National Library of Wales. The time had come for George Fisher's departure. The longest serving member of the Press staff, he had presided over the bindery for more than twenty years. Of the senior staff he appears to have been the only one consistently happy at Gregynog and the only one to leave without bitterness. His long and devoted service was marked by the presentation of a cheque for one hundred guineas.

As the horror of war receded and things gradually returned to normality, Gwen's thoughts turned towards the possible re-opening of the Press, though Burdon Evans, Colonel Powell and Daisy were all opposed to the idea. Wardrop, now back at the Victoria and Albert Museum, was willing to resume his former duties, and Gwen thought it would be possible to start again in a modest way. In May 1945, she once again turned to her old friend, T. J., for support:

'I do hope that you will back me up in *one* thing – the continuance, if at all possible, of the Gregynog Press . . . only you and I care anything about the Press, as you know.'[9]

This time her sister was not to be invited to join in the enterprise in which she had long ago lost all interest.

It was some time now since the last wounded serviceman had left Gregynog Hall. In the summer and early autumn of 1945, two batches of Dutch children were invited to stay for several weeks. These were followed by a third contingent in March 1946 and a fourth the following May.

In February 1946, the Press Board met formally to discuss the possible re-opening of the Press, but as had been forecast in 1939, it proved impossible to reassemble the old team. Fisher had retired to his farm in Hampshire, his assistant, John Bowen, was negotiating for a post at the National Library of Wales, where Idris Jones had already secured employment as a printer. R. A. Jones, a bindery apprentice taken on shortly before the war, had been engaged by W. T. Morrell Limited of London to finish his apprenticeship, and J. O. Morgans, compositor from January 1939 until March 1940, had joined the staff of Sellar and Sellar of Hayes, Middlesex. Though tentative suggestions were put forward for producing a modest series of Welsh books, the odds were stacked so heavily against a re-opening, that finally the idea had to be abandoned.

When all hope was fled, Gwen Davies penned T. J. a touching letter of appreciation:

'It is just to tell you that you have been the big thing in my life. I was so hungry for affection and encouragement: the last war [1914–1918] had knocked my life and health to pieces, and it all had to be rebuilt, reconstructed anew, and you were the builder and restorer. We planned and worked together. I do want to thank you from the bottom of my heart for all you have given me and done for me. It has been like a warm fire glowing within me for years. It has just raised my life out of the humdrum rut of existence until I felt I could tackle anything with your help and co-operation. All I have been able to do during the last thirty years has been almost entirely your doing. Thank you.'[10]

At T. J.'s request, Colonel Powell prepared a financial statement in August in which it was revealed that the total loss on the Press between April 1927 and March 1940 amounted to £6,200. This took no account of the £1,000 per annum settled on the Press by the sisters. Sales of books during wartime had remained surprisingly reassuring and the numbers were recorded as follows:

1939–1940	..	219
1940–1941	..	346
1941–1942	..	315
1942–1943	..	420
1943–1944	..	536
1944–1945	..	209

Copies of *The Plays of Euripides* and *The Lamentations of Jeremiah* alone still remained unsold.

By the end of 1946, Gwen Davies, who had been staying with friends in Scotland, became seriously ill and spent several weeks undergoing a series of blood transfusions in Edinburgh Royal Infirmary. A long convalescence followed, and she remained in Scotland until July 1947. Her homecoming was saddened by the recent death of Jane Blaker, Hugh Blaker's sister and their friend and companion for more than fifty years, who had died of meningeal thrombosis. During the latter half of the year Gwen's health improved, but by February 1948 she had to endure another series of blood transfusions in Cardiff.

In 1949, Daisy Davies received an honorary doctorate of laws of the University of Wales. Meanwhile, in the New Year's Honours List W. J. Burdon Evans had been made a Commander of the Order of the British Empire, an award he had not long to enjoy for he died the following March. Yet another friend and confidant was removed from the thinning Davies circle. Nine months later, in January 1950, Thomas W. Hughes who had served faithfully, if somewhat controversially, as estate agent since the early years at Gregynog, also died.

For the last four years of her life Gwen Davies lived quietly, spending the greater part of her time at Gregynog, much of it in a wheelchair. Her last extant letter to T. J. is dated 11 June 1951. On 24 June she left Gregynog Hall for the last time and entered the Radcliffe Infirmary, Oxford. She died there on 3 July and was cremated the following day in Oxford.[11] The cause of death was recorded as aplastic anaemia, the blood disease which had first been diagnosed in France during the First World War and which had affected her health ever since.

Idris Jones printed at the National Library of Wales the orders of two memorial services, one held on Friday, 7 September, in the Music Room at Gregynog and one on the following Sunday at the Presbyterian chapel in Llandinam. A great congregation of old friends gathered on the former occasion. George Fisher, who had been invited by T. J., wrote regretting that he would not be able to attend but that,

'I very much wish I could help in paying that last tribute to a great and quiet life work.'[12]

The service on this occasion was arranged by Dora Herbert Jones as a mark of affection to the friend from whom she had latterly been almost estranged. (Four items included in the service – *Psalms XXIII* and *CXXI*, Walford Davies's setting of *Lord, It Belongs Not to My Care* and a reading of Henry Vaughan's poem, *They Are All Gone Into the World of Light* – were to be used nearly twenty-three years later for Dora's own memorial service, held in the same room. In spirit the friends were thus finally reunited in the place they both loved.)[13] After the service, T. J. wrote to Dora,

'I am so grateful to you for all you meant to us yesterday – the help to voice our feelings in fitting words and music. No two persons in that congregation of friends had been nearer to her heart and mind than you and me in her most creative period. She was far from flawless, so are we, and I most certainly; but she had to carry a burden of heredity and environment which you and I have been spared.'[14]

The letter went on to ask for Dora's help in preparing a short history of the Press, but, perhaps understandably, she felt that the close emotional involvement which had been hers was not conducive to objectivity, and she declined to write anything.

At Gwen's death her estate amounted to £863,662. In her will, she bequeathed seventy-eight oil paintings, twenty-six drawings and watercolours and five pieces of sculpture to the National Museum of Wales in Cardiff. The paintings, which included four Corots, five Millets, two Manets, two Renoirs, six Monets, three Cezannes, seven Daumiers, three Whistlers, three Gainsboroughs, seven Turners and a Constable, were to transform the Cardiff gallery from a small provincial collection into one of national importance. The official opening, presided over by Sir Leonard Twiston-Davies, assisted by the Director, Dr Dilwyn John, was performed by the Duchess of Kent on 27 May 1952.[15]

After her sister's death, Daisy lived quietly at Gregynog and continued to purchase paintings. These, though interesting, were not quite in the same class as those bought by the sisters between 1908 and 1924. During the years 1955 to 1961, she also revived a modest version of the old summer festivals, held under the direction of Ian Parrott, Gregynog Professor of Music at University College Aberystwyth, but these one-day events, despite the appearance of a few eminent musicians, were but pale shadows of their predecessors. Occasional conferences too continued to be held.

So far, the plant and equipment had remained untouched in the Press, since Gwen Davies had stoutly refused to part with any of it. After her death, George Fisher tried to acquire some of the binding tools and equipment for his own use in Hampshire in return for finishing the sixteen remaining copies of *The Plays of Euripides*, but though T. J. interceded with Daisy on his behalf, she refused to accede to the request.

On 7 April 1954, T. J. addressed fellow members of The Double Crown Club[16] at a dinner held at Kettner's Restaurant, Romilly Street, London. Sir Francis Meynell presided, and the sixty or so diners included Blair Hughes-Stanton and James Wardrop. The subject of his after-dinner paper was the Gregynog Press. This brief outline history, which he had prepared with the help of notes from ex-members of the Press staff, was expanded and a bibliography of the books appended for the version published by the Oxford University Press later in the year.[17] Despite some inaccuracies and the brevity of the text, this has remained to date a standard source book on the Press.

The same month, Daisy Davies decided to donate the bulk of the printing and binding equipment to the National Library of Wales. Idris Jones was sent over to supervise the packing of the Monotype caster, the platen and hand presses, together with type matrices, some of the type, all the wood engraved blocks and the binding tools. Some of the less valuable pieces of equipment were left in the Press at Gregynog where they still remain.

After resigning his presidency of University College Aberystwyth, T. J. moved in November 1954 to St Nicholas-at-Wade, Essex, where he had originally built a cottage in the 1920s. Always a copious letter writer, he continued to correspond with a vast circle of friends, including Daisy Davies.[18] In June 1955 he suffered severe injuries as a result of a fall in his home. From these he never recovered and he died in London on 14 October. His long friendship with the Davies sisters, his contribution to the Gregynog venture in general and to the success of the Press in particular were commemorated in a service held in the Music Room at Gregynog on 10 December 1955. It has been remarked, on more than one occasion and with some justification, that the Davies's may have supplied the money but that it was T. J.'s press. Certainly without him there could have been no press. He it was who forged all the essential links – literary, artistic and craft – who chaired the Press Board and made most of the major decisions, who smoothed the path in difficult situations, who gloried with Gwen Davies in moments of triumph and who supported her unfailingly in times of trouble.

Daisy Davies, now well advanced in years, turned her attention to Gregynog's future. After long deliberation and consultation, in 1960 she gave the house and grounds to the University of Wales as a venue for inter-collegiate courses, conferences and seminars. She continued to live there, as a tenant of the University, and the house remained as it was until after her death, which occurred on 14 March 1963. Since that time, certain alterations and extensions have been made to the house so as to allow accommodation for quite large groups. Here students now study while taking delight in their gracious surroundings. Gregynog's future as a cultural centre is assured.

In 1974, the University decided to attempt to revive the Press on a modest scale, under the imprint *Gwasg Gregynog*, and in September appointed Michael Hutchins as the first Printing Fellow to take charge of the venture. The first book, R. S. Thomas's *Laboratories of the Spirit*, printed in Janson type on hand-made paper, was published in 1976. One wishes the experiment well and can only regret that the days of printing several de luxe editions per year are over. Institutional funds and corporate imagination can never, alas, encompass vistas and standards forever fled.

NOTES

1. Letter from T. J. to Gwen Davies, dated 27 December 1939.
2. A reference to *The Praise and Happinesse of the Countrie-Life.*
3. Letter from James Wardrop to T. J., dated 28 December 1939.
4. Letter from Gwen Davies to T. J., dated 18 November 1939.
5. The story of this incident is told in Meta Maclean's *The Singing Ship: an Odyssey of Evacuee Children*, Angus and Robertson, 1941.
6. Letter from James Wardrop to T. J., dated December 1941.
7. Letter from Gwen Davies to T. J., undated.
8. Letter from Gwen Davies to T. J., dated 16 March 1942.
9. Letter from Gwen Davies to T. J., dated 4 May 1945.
10. Letter from Gwen Davies to T. J., dated 12 August 1946.
11. An obituary notice, written by T. J., appeared in *The Observer*, 8 July 1951.
12. Letter from George Fisher to T. J., dated 24 July 1951.
13. Dora Herbert Jones died 9 January 1974.
14. Letter from T. J. to Dora Herbert Jones, dated 8 September 1951.
15. The collection was further augmented by bequests after Daisy Davies's death in 1963.
16. A dining club founded in 1924 by Holbrook Jackson, Oliver Simon and Hubert Foss and which has included among its members the most distinguished printers, typographers and bibliophiles.
17. Thomas Jones. *The Gregynog Press: a paper read to the Double Crown Club on 7 April 1954.* Oxford University Press, 1954. Edition limited to 750 copies. Chosen as one of the fifty exhibits in the National Book League's *Book Design Exhibition of the Best Books of 1954.*
18. T. J.'s last letter to Daisy Davies is dated 26 July 1955.

Engraving by Reynolds Stone from
The Praise and Happinesse of the Countrie-Life.

Bibliography of the Books Printed at the Press

THE ARRANGEMENT IS CHRONOLOGICAL EXCEPT for the three books printed for private circulation which appear at the end. Capitalisation has been standardised and kept to a minimum, but the punctuation is as it appears on the title-page. Publication dates are taken from the bibliography in Thomas Jones's *The Gregynog Press: a Paper Read to the Double Crown Club on 7 April 1954*, the only source of information extant, except in two cases where the dates quoted are obviously inaccurate. The controller of the Press is normally cited as printer in the colophon. Colophons are not transcribed as many are of inordinate length. The precise measurement of uncut leaves within a copy of a title is subject to variation. The majority of the ordinary bindings were case-bound by hand. Where a binding is described as paper covered, this indicates a paper covering over boards. Except where otherwise described, all have endpapers to match the text paper. Pages are uncut or trimmed at the top edge only. The special bindings have endpapers to match the text paper unless otherwise described. All headbands are hand-sewn. The leaves are uncut at the fore-edge and tail, and the top edges are gilt. Where the signature includes a name other than that of the binder it is that of the designer of the binding. As many variants as possible have been checked, where locations are known.

1 1923 **Poems by George Herbert**

[in red] Poems [in black] by George Herbert [in red Device No. 7] The Gregynog Press MCMXXIII

xv [1] 26 [2] pp. Published 7 December 1923. Kennerley type on Grosvenor Chater hand-made paper. 1 wood engraving and wood engraved initial letters by R. A. Maynard. Printed on a hand-press. Assistants: John Mason and John Hugh Jones; Poems selected by H. Walford Davies; Completed on the ninth day of November, 1923 (colophon). 202 × 140 mm. 257 copies bound in grey Gregynog marbled paper with natural cloth spine. Printed title label on upper cover. Signed

on printed label inside front cover BOUND AT THE GREGYNOG PRESS. Price ½ guinea.
Special binding: 205 × 142 mm. 43 copies, the majority bound in crimson levant morocco, polished. Covers plain except for gilt rule frame. 5 raised bands with gilt spine titling. Board edges gilt ruled. 7 oblique gilt pallet lines on headcaps. Red silk headbands. Turn-ins ruled in gilt. Pink and blue marbled endpapers. Unsigned but 30 copies bound by John Mason, 13 copies bound later by George Fisher. Price 1½ guineas. Variants bound in blue levant morocco, niger or vellum.

2 1924 **Poems by Henry Vaughan**
[in red] Poems [in black] by Henry Vaughan [in red Device No. 7] The Gregynog Press MCMXXIV
xxxv [2] 2–85 [3] pp. Published December 1924. Kennerley type on Grosvenor Chater hand-made paper. 13 wood engravings by R. A. Maynard and H. W. Bray, wood engraved initials by R. A. Maynard. Assistants: John Mason and John Hugh Jones; The poems were selected by Ernest Rhys; Completed the fourth day of December 1924 (colophon). 210 × 145 mm. 470 copies bound in grey Ingres paper, blocked in blue with the swan of Usk designed by H. W. Bray and an HV monogram flanked on both sides by the letter S (Henry Vaughan Silurist) and the dates 1622–1695. Dark blue cloth spine with title and imprint in gilt. Unsigned. Price 1½ guineas.
Special binding: 205 × 140 mm. First 30 copies bound in crimson levant morocco, polished. Plain except for swan of Usk motif blocked in gilt on upper cover. 5 raised bands outlined in blind with gilt spine titling. Board edges gilt ruled. 5 oblique gilt pallct lines on headcaps. Buff silk headbands. Turn-ins decorated with double gilt rule. Blue and pink marbled endpapers. Unsigned. Price 3 guineas. Variants bound in blue levant morocco.

3 1925 **Caneuon Ceiriog detholiad**
Caneuon [in red] Ceiriog [in black] detholiad [in red Device No. 13] Gwasg Gregynog MCMXXV
xxx [2] 87 [1] pp. Published 22 May 1925. Kennerley type on Grosvenor Chater hand-made paper. Text in Welsh, printed in red and black. Wood engraved frontispiece and 30 other wood engravings by R. A. Maynard and H. W. Bray. Compositor: John Hugh Jones; Pressman: John Mason; Poems selected and edited by John Lloyd Jones; Completed 7 February

1925 (colophon). 260 × 200 mm. 370 copies in grey paste-marbled paper with natural cloth spine and printed spine label. Unsigned. Price 1 guinea. *Special binding:* 257 × 200 mm. First 30 copies bound in red levant morocco, polished. Covers richly tooled in gilt with pattern of concentric panels and corner leaf sprays surrounding a plain central panel, outlined in gilt with corner fleurons. 4 raised bands, gilt panelled spine bearing title and leaf sprays in gilt. Buff silk headbands. Turn-ins decorated with 4 gilt rules and corner fleurons. Red, turquoise and grey marbled endpapers. Signed in gilt on turn-in of lower cover THE GREGYNOG PRESS BINDERY 1926. Price 3½ guineas. Variants bound in blue levant morocco.

4 1926 **Detholiad o ganiadau gan T. Gwynn Jones**

Detholiad o ganiadau gan T. Gwynn Jones [in grey Device No. 4] Gwasg Gregynog MCMXXVI

xiii [3] 169 pp. Published ? Thomas Jones's bibliography cites 8 January 1926 which is clearly an error. Kennerley type on Dutch hand-made paper. Text in Welsh. 9 head- and tail-pieces engraved on wood by R. A. Maynard. Compositor: John Hugh Jones; Pressmen: H. W. Bray and Idris Jones; Completed 7 May 1926 (colophon). 210 × 142 mm. 474 copies bound in peacock blue buckram with gilt spine titling. Unsigned. Price 15*s*.

Special binding: 210 × 142 mm. 26 copies (Nos. 1–25 and No. 30) bound in dark blue levant morocco, polished. Covers decorated with a trellis of double blind lines which form squares, with small gilt circles at intersections, surrounded by a gilt rule from the spine edge of which springs a pattern of gilt lines, terminating in small gilt circles. 5 raised bands outlined with double and single gilt rules. Gilt titling. Red silk headbands. Turn-ins decorated with 1 gilt and 1 double blind rule which has small gilt circles at outer corners. Blue-grey Ingres endpapers. Signed in gilt on turn-in of lower cover GREGYNOG PRESS BINDERY. Binding design by H. W. Bray. Price 2½ guineas. Variants bound in green or blue levant morocco.

5 1926 **Chosen essays by Edward Thomas**

Chosen essays by Edward Thomas [wood engraved portrait within a double rule frame] The Gregynog Press MCMXXVI

102 [2] pp. Published 10 November 1926. Garamond type. The first 50 copies on Batchelor's hand-made paper, watermarked GG, 300 copies on

Van Gelder hand-made paper. 24 wood engravings by R. A. Maynard and H. W. Bray. Essays selected by Ernest Rhys. Compositors and assistants: Idris Jones, John Hugh Jones and Richard Owen Jones; Completed on the seventh day of August 1926 (colophon). Within the colophon, Device No. 4 in turquoise. 250×190 mm. 317 copies bound in buckram, the majority in blue but a few copies in yellow, blocked in gilt on the upper cover with an outline vase of flowers. Gilt spine titling. Unsigned. Price £1.5*s*.

Special binding: 250×190 mm. 33 copies (Nos. 1–30, 32, 49 and 50) bound in blue morocco, polished. Covers decorated in gilt with two sets of concentric squares, set one above the other, and a trellis of double blind lines which have small gilt squares at the intersections. 5 raised bands outlined in blind. Gilt titling and panels outlined in gilt. 2 gilt squares at head and tail enclosed in blind lines. Board edges ruled in gilt. Pale green silk headbands. Turn-ins decorated with double gilt rule. Blue-grey silurian endpapers. Signed in gilt on turn-in of lower cover GREGYNOG PRESS BINDERY. Binding design by R. A. Maynard. Price 4½ guineas.

6 1927 **Selected poems of Edward Thomas**

Selected poems of Edward Thomas [wood engraving in blue] With an introduction by Edward Garnett [in blue] The Gregynog Press MCMXXVII

[iv] v–xix [1] 95 [1] pp. Published 8 August 1927. Garamond type on Japanese vellum. Wood engraved shadow initial letters in red. Pages ruled in blue. On half-title – Selected poems *by* Edward Thomas. Assistants: H. W. Bray, George Fisher; Compositor: Richard Owen Jones; Pressmen: John Hugh Jones and Idris Jones; Completed 24 June 1927 (colophon). Within the colophon, Device No. 7 in red. 232×153 mm. 250 copies bound in yellow buckram with gilt spine titling. Unsigned. Price £1. 5*s*.

Special binding: 228×150 mm. First 25 copies bound in blue levant morocco, polished. Covers tooled in gilt with a pattern of leaf sprays arranged as overlapping lozenge shapes which are connected by further decoration to a double gilt rule frame. 5 raised bands outlined in blind. Gilt spine titling. First and sixth spine panels tooled in gilt with a leaf spray. Double gilt rules at head and tail. Gilt ruled board edges. Head-caps tooled with three gold dots. Turn-ins decorated with double gilt rule. Blue Cockerell marbled endpapers. Signed in gilt on turn-in of lower cover R. ASHWIN MAYNARD GEORGE FISHER GREGYNOG PRESS BINDERY. Price 7 guineas.

7 1927 The life of Saint David

[in blue, wood engraved title] The life of Saint David [wood engraving] MCMXXVII The Gregynog Press [blue cross]

41 [3] pp. Published 24 November 1927. Poliphilus type on Batchelor hand-made paper. 25 wood engravings by R. A. Maynard and H. W. Bray, hand coloured. Chapter openings in pale blue, initial letters in red and paragraph marks in red, drawn with a quill by H. W Bray. Compositor: Richard Owen Jones; Pressmen: John Hugh Jones and Herbert John Hodgson; Completed 27 September 1927 (colophon). Device No. 7 in red follows the colophon. 250 × 190 mm. 150 copies bound in limp vellum with gilt spine titling. Unsigned. Price 5 guineas.

Special binding: 250 × 190 mm. First 25 copies bound in scarlet levant morocco, polished. Covers blocked in gilt with large outline cross framed with double gilt rule. 5 raised bands. Spine panelled in gilt with gilt titling. Board edges ruled in gilt. 3 gilt dots on headcaps. Turn-ins decorated in gilt with triple rule fillet. Signed in gilt on turn-in of lower cover HORACE W. BRAY GEORGE FISHER GREGYNOG PRESS BINDERY. Price 12 guineas.

8 1927 Llyfr y pregeth-wr

[in red, wood engraved title] Llyfr y pregeth-wr [wood engraving] Gwasg Gregynog MCMXXVII

20 [2] pp. Published 24 November 1927. Poliphilus type in red and black on Batchelor hand-made paper. Wood engraved frontispiece and title-page wood engraving by David Jones. Compositor: John Hugh Jones; Pressman: Herbert John Hodgson; Completed 1 October 1927 (colophon). Device No. 7 in red follows colophon. 256 × 195 mm. 223 copies bound in limp blue buckram with titling in gilt on the upper cover. Unsigned. Price 12*s.* 6*d.*

Special binding: 253 × 192 mm. First 25 copies bound in dark blue levant morocco, polished. Gilt rule frame to covers. 5 raised bands. Spine panelled in gilt with gilt titling. Board edges ruled in gilt. Blue silk headbands. Turn-ins decorated in gilt with double rule fillet. Signed in gilt on turn-in of lower cover GREGYNOG PRESS BINDERY. Price 3½ guineas.

Nos. 203 and 249 printed on vellum and bound in blue levant morocco to a design by George Fisher, decorated with coloured onlays and gilt and blind tooling.

9 1928 **An account of the convincement of . . . Richard Davies**
An account of the convincement, exercises, services and travels of that ancient servant of the Lord [double rule] Richard Davies [double rule] with some relation of ancient friends and the spreading of truth in North Wales & c [in red Device No. 7] The Gregynog Press [double rule] MCMXXVIII
xx [2] 166 [2] pp. Published 2 April 1928. Baskerville type on Batchelor hand-made paper. Compositor: Richard Owen Jones; Pressman: Herbert John Hodgson; Completed 7 February 1928 (colophon). 215 × 140 mm. 150 copies bound in dark blue buckram titled in gilt on spine and upper cover. Unsigned. Price 2 guineas.
Special binding: 215 × 140 mm. First 25 copies bound in dark blue levant morocco, polished. Covers plain. 5 raised bands tied down with blind ties. Gilt spine titling. Board edges ruled in gilt. 3 gilt dots on headcaps. Pale blue silk headbands. Turn-ins decorated in gilt with double rule fillet. Signed in gilt on turn-in of lower cover GEORGE FISHER GREGYNOG PRESS BINDERY. Price 5 guineas.

10 1928 **The autobiography of Edward Lord Herbert of Cherbury**
The autobiography of Edward Lord Herbert of Cherbury with an introduction by C. H. Herford [ornament] [wood engraved shield hand coloured in blue, grey and red] The Gregynog Press MCMXXVIII [ornament]
[vi] vii–xv [3] 93 [3] pp. Published 28 May 1928. Poliphilus and Blado italic on Batchelor hand-made paper. 9 wood engravings by H. W. Bray. Large opening wood engraved initial I in red. Compositor: Richard Owen Jones; Pressmen: John Hugh Jones and Herbert John Hodgson; Completed 10 March 1928 (colophon). Within the colophon, Device No. 7 in red. 340 × 235 mm. 275 copies bound in brown buckram with Herbert arms blocked in gilt on upper cover. Gilt spine titling. Unsigned. Price 4 guineas.
Special binding: 338 × 232 mm. First 25 copies bound in brown levant, polished. Blocked in palladium on upper cover with Herbert arms on red and blue onlays. Shield outlined in heavy gilt. 5 raised bands outlined in blind. Gilt spine titling. Scarlet silk headbands. Turn-ins decorated in gilt with heavy double fillet rule. Signed in gilt on turn-in

of lower cover GREGYNOG PRESS BINDERY GEORGE FISHER. Price 12 guineas.

11 1928 Selected poems of W. H. Davies

Selected poems of W. H. Davies arranged by Edward Garnett with a foreword by the author [in yellow and black Device No. 9] The Gregynog Press MCMXXVIII

[iv] v–vii [1] 91 [1] pp. Published 9 November 1928. Baskerville type on Japanese vellum. Wood engraved frontispiece portrait of the author by R. A Maynard after the portrait by Augustus John. Pages ruled in red. Assistant: H. W. Bray; Compositor: Idris Vaughan Jones; Pressman: Herbert John Hodgson; Completed 25 October 1928 (colophon). 230×155 mm. 285 copies bound in black Cockerell marbled paper with black buckram spine, titled in gilt, and black buckram fore-edge strip. Unsigned. Price 1½ guineas.

Special binding: 227×150 mm. First 25 copies bound in sage green morocco, polished. Each cover tooled in gilt with three large foliate devices enclosed in two bands of seven lines which continue across the spine. At head and tail are two further bands of four lines which also continue across the spine. 5 raised bands, the first and fifth of which are outlined in blind, the second and fourth outlined by bands of gilt lines. Gilt spine titling. Gold silk headbands. Turn-ins decorated with gilt lines continuing from the covers. Signed in gilt on turn-in of lower cover R. ASHWIN MAYNARD GREGYNOG PRESS BINDERY GEORGE FISHER. Price 8½ guineas.

12 1928 The misfortunes of Elphin

The misfortunes of Elphin by Thomas Love Peacock [quotation] [wood engraving] The Gregynog Press 1928

119 [1] pp. Published 23 November 1928. Baskerville type on Batchelor hand-made paper. 21 wood engravings, forming head- and tail-pieces, by H. W. Bray. Compositor: Richard Owen Jones; Pressman: Herbert John Hodgson; Completed 9 October 1928 (colophon). Device No. 8 in black precedes the colophon. 240×160 mm. 225 copies bound in patterned cloth, blue and black, with purple buckram spine titled in gilt. Unsigned. Price 2 guineas.

Special binding: 240×156 mm. First 25 copies bound in red levant morocco, polished. Upper cover blocked in gilt with a design of two hand-held goblets in which wine is being poured from one to the other, surrounded by a gilt rule frame. 5 raised bands. Spine gilt panelled with gilt titling. Board edges ruled in gilt. 3 gilt dots on headcaps. Red silk headbands. Turn-ins decorated with two fine gilt rules. Signed in gilt on turn-in of lower cover HORACE W. BRAY GREGYNOG PRESS BINDERY GEORGE FISHER. Price 7 guineas.

13 1928 **Penillion Omar Khayyâm**

[in blue] Penillion [in black] Omar Khayyâm Wedi eu cyfeithu o'r Berseg i'r Gymraeg gan John Morris Jones [in blue and black Device No. 9] Gwasg Gregynnog MCMXXVIII

[iv] v–xvi [1] 2–30 pp. Published 7 December 1928. Caslon type on Batchelor special making of GG watermarked hand-made paper. Text printed in blue and black. 10 wood engravings by R. A. Maynard. Compositor: John Hugh Jones; Pressman: Herbert John Hodgson; Completed 6 November 1928 (colophon). 245 × 175 mm. 285 copies bound in white buckram with yellow buckram spine, titled in gilt, and yellow buckram fore-edge strip. Unsigned. Price 15*s*.

Special binding: 247 × 174 mm. First 25 copies bound in blue levant morocco, polished. Covers plain. 5 raised bands outlined in blind. Gilt spine titling. Blue silk headbands. Turn-ins decorated in gilt with double rule fillet. Signed in gilt on turn-in of lower cover GEORGE FISHER GREGYNOG PRESS BINDERY. Price 4 guineas.

14 1929 **Psalmau Dafydd**

[wood engraved title-page in red and black] Psalmau Dafydd yn ol William Morgan 1588 Gwasg Gregynnog 1929

89 [3] pp. Published 30 October 1929. Poliphilus type on Batchelor hand-made paper. Title-page, in which the title is framed in a decorative foliate border which encloses a unique form of the Press device, decorative wood engraved openings and initials designed and cut by H. W. Bray. Text printed in red, blue and black. Text edited by Ifor Williams. Compositors: John Hugh Jones and Idris Vaughan Jones; Pressman: Herbert John Hodgson; Completed 10 April 1929 (colophon). 290 × 210 mm. 200 copies bound in a paper printed in red on buff from a design engraved on wood at the Press. Red niger morocco spine and red buckram corner pieces. 5 raised bands outlined in blind. Gilt spine titling. Blue silk headbands. Top edge gilt, fore-edge and tail uncut. Signed on paste-down of lower cover GREGYNOG PRESS BINDERY. Price 5 guineas.

Special binding: 288 × 210 mm. First 25 copies bound in blue levant morocco, polished. Covers plain. 5 raised bands outlined in blind. Titling, cross and date in gilt. Pale blue silk headbands. Turn-ins tooled in gilt with 6-rule and 4-rule fillets. Signed in gilt on turn-in of lower cover GREGYNOG PRESS BINDERY GEORGE FISHER. Price 9 guineas.

15 1930 **Christina Rossetti: Poems**

Christina Rossetti [swelled rule] Poems chosen by Walter de la Mare [Device No. 12] The Gregynog Press MCMXXX

xliii [1] 107 [1] pp. Published 7 April 1930. Baskerville type on Japanese vellum. Wood engraved portrait and shadow initial letters by R. A. Maynard. Hanging titles in red. Assistant: H. W. Bray; Compositor: John Hugh Jones; Pressman: Herbert John Hodgson; Completed 21 January 1930 (colophon). 232 × 152 mm. 275 copies bound in grey and white Cockerell marbled paper. Natural Hermitage calf spine. Author in gilt within gilt panel which continues to covers. Signed on paste-down of lower cover GREGYNOG PRESS BINDERY. Price 2½ guineas.

Special binding: 230 × 150 mm. First 25 copies bound in scarlet levant morocco, unpolished. Covers tooled in gilt with 2 large foliate ornaments combined with diagonal blind decoration and gold dots, set one above the other, and surrounded with more gold dots. Blind tooled frame, pierced by blind lines and gold dots, is bordered by a single rule gilt frame. 5 raised bands. Spine panels outlined in blind. Author's name in gilt. Board edges ruled in gilt. 3 gilt dots on headcaps. Turn-ins tooled in gilt with double rule fillet. Signed in gilt on turn-in of lower cover R. ASHWIN MAYNARD GEORGE FISHER GREGYNOG PRESS BINDERY. Price 8½ guineas.

16 1930 **The stealing of the mare**

The celebrated romance of the stealing of the mare translated from the original Arabic by Lady Anne Blunt and done into verse by Wilfrid Scawen Blunt [in gilt with circular grey-blue surround Device No. 13] The Gregynog Press 1930

[vi] vii–ix [3] 73 pp. Published 22 November 1930. Garamond type on heavy Japanese vellum. Wood engraved opening and 13 decorated initials by R. A. Maynard, hand coloured. Compositor: Richard Owen Jones; Pressman: Herbert John Hodgson; Completed 1 July 1930 (colophon). 305 × 227 mm. 250 copies bound in paper printed in three colours, designed and printed at the Press. Natural Hermitage calf

spine with gilt titling. Signed on paste-down of lower cover GREGYNOG PRESS BINDERY. Price 4 guineas.

Special binding: 305 × 225 mm. The first 25 copies bound in green levant morocco, unpolished. Covers plain except for gilt rule frame. 5 raised bands. Spine panels outlined in gilt. Gilt titling. Board edges ruled in gilt. 3 gilt dots on headcaps. Light blue silk headbands. Turn-ins tooled in gilt with double rule fillet. Signed in gilt on turn-in of lower cover GREGYNOG PRESS BINDERY GEORGE FISHER. Price 8 guineas.

17 1931 Elia

Elia and the last essays of Elia by Charles Lamb [double rule] in two volumes With wood-engravings adapted from contemporary prints Vol. I Elia [Device No. 11] [double rule] The Gregynog Press MCMXXIX

[iv] v–vii [1] 265 [3] pp.

Elia and the last essays of Elia by Charles Lamb [double rule] in two volumes With wood-engravings adapted from contemporary prints Vol. II The last essays of Elia [Device No. 11] [double rule] The Gregynog Press MCMXXX

[iv] v–vii [1] 219 [3] pp. Published 19 February 1931. Baskerville type on Japanese vellum. Vol. 1 has 16 and Vol. 2 11 wood engraved head-pieces by H. W. Bray adapted from prints in the collection of Edward J. Finch. Compositors: Richard Owen Jones, John Hugh Jones, Idris Vaughan Jones; Pressman: Herbert John Hodgson; Vol. 1 completed 14 November 1929 and Vol. 2 completed 22 March 1930 (colophon). 230 × 155 mm. 260 copies bound in grey buckram with gilt spine titling. Unsigned. Price 5 guineas.

Special binding: 230 × 150 mm. First 25 copies bound in brown levant morocco, unpolished. Covers plain except for blind rule frame which is carried over the spine. 5 raised bands tied down with blind ties. Gilt spine titling. Light brown silk headbands. Turn-ins tooled with double rule fillet. Signed in gilt on turn-in of lower cover GEORGE FISHER GREGYNOG PRESS BINDERY. Price 10 guineas.

18 1931 The plays of Euripides

The plays of Euripides [double rule] Volume I [in red] Hippolytus The Bacchae The Trojan women Electra [in black] translated into English rhyming verse by Gilbert Murray, LL.D., D.Litt. with wood-engravings from the Greek vase paintings by Robert Ashwin Maynard and Horace Walter Bray [in red Device No. 7] The Gregynog Press [double rule] MCMXXXI

ix [3] 269 [1] pp.

The plays of Euripides [double rule] Volume II [in red] Medea The Iphigenia in Taurus Alcestis The Rhesus [in black] translated into English rhyming verse by Gilbert Murray, LL.D., D.Litt. with wood-engravings from the Greek vase paintings by Robert Ashwin Maynard and Horace Walter Bray [in red Device No. 7] The Gregynog Press [double rule] MCMXXXI

[5] 7–263 [1] pp. Published 20 July 1931. Bembo and Fairbank's italic type on Batchelor special making of GG watermarked hand-made paper. Each volume has 16 engravings designed by R. A. Maynard and H. W. Bray and cut on wood by H. W. Bray, the subjects chosen by T. B. L. Webster from 5th century Greek vase paintings. Compositors: John Hugh Jones, Idris Vaughan Jones and Richard Owen Jones; Pressman: Herbert John Hodgson; Vol. 1 completed 17 March 1931 and Vol. 2 completed 27 June 1931; One hundred and twenty-five copies reserved for Random House, New York (colophon). 378 × 250 mm. 475 copies bound in brown Sundour linen cloth blocked in gilt on the upper cover with a Press device based on Device No. 12 and gilt spine titling. Unsigned. Price 12 guineas.

Special binding: 375 × 245 mm. First 9 copies bound as one volume in brown levant morocco, polished. Blocked in gilt on both covers with the figure of Apollo enclosed by bands of fine gilt lines. 5 raised bands. Spine tooled in gilt with title, imprint and bands of fine lines. Brown silk headbands. Turn-ins tooled in gilt with 6-rule fillet. Signed in gilt on turn-in of lower cover GEORGE FISHER GREGYNOG PRESS BINDERY. Price 20 guineas. Copies 10–25 bound in 1952 by John Ewart Bowen at the National Library of Wales in brown levant morocco, polished. Design similar to the other 9 copies except that the bordering bands of fine lines are replaced by a triple rule fillet.

19 1931 **Comus**

[wood engraved title] Comus a mask by John Milton with a frontispiece and the six characters in costume designed and engraved on wood by Blair Hughes-Stanton [wood engraving] The Gregynog Press MCMXXXI

[10] 25 [3] pp. Published 2 October 1931. Baskerville type on Japanese vellum. The first book arranged by William MacCance and completed

on the hand-press on the eleventh day of June, 1931 . . . Blair Hughes-Stanton collaborated in the printing . . . The book was set by hand and printed by Idris Jones (colophon). 277 × 173 mm. 225 copies bound in buff paper with yellow buckram spine strip, titled in gilt on spine and front. Unsigned. Price £1. 15*s.*
Special binding: 275 × 172 mm. First 25 copies bound in dun-coloured levant morocco, polished. Covers decorated with two vertical panels of blind lines, bordered by light and heavy gilt frames. On the upper cover the inner panel is interrupted by a panel containing the title in blind. 5 raised bands. Spine panels outlined with double gilt frames. Blind spine titling. Buff silk headbands. Turn-ins tooled with a broad blind rule and finer gilt rule which is continued from the covers. Signed on turn-in of upper cover in blind GREGYNOG PRESS BINDERY and in blind on turn-in of lower cover BLAIR HUGHES-STANTON GEORGE FISHER. Price 8 guineas.

20 1932 **The fables of Esope**
[wood engraved title] The fables of Esope [wood engraving] translated out of Frensshe in to Englysshe by: William Caxton with engravings on wood by Agnes Miller Parker [wood engraving] The Gregynog Press: Newtown Montgomeryshire:—MCMXXXI
[8] 146 [2] pp. Published 2 April 1932. Bembo type on Barcham Green hand-made paper. 37 wood engravings by Agnes Miller Parker and wood engraved initial letters by William MacCance. Compositor: Richard Owen Jones; Pressman: Herbert John Hodgson; In this edition obvious errors in the original text have been corrected. The diagonal dash / used by Caxton was found to be disturbing with the type adopted and was replaced by the colon : Completed 30 November 1931 (colophon). Within the colophon, Device No. 18 in black. 308 × 210 mm. 225 copies bound in Welsh natural sheepskin, tanned locally. Spine titling blocked in black. Signed on paste-down of lower cover GREGYNOG PRESS BINDERY. Price 5 guineas.
Special binding: 305 × 215 mm. First 25 copies bound in dun-coloured levant morocco, polished. Upper cover decorated with an ingenious arrangement of the letters forming ESOPE onlaid in reddish-brown Hermitage calf, the remainder of the title tooled in blind and gilt. Surrounding gilt rule frames continue to the lower cover where they enclose a press device consisting of shadow gilt letters tooled on reddish-brown onlay. 5 raised bands outlined in gilt. Titling and date tooled in

gilt and the author's name in blind, combined with 8 small squares, onlaid in reddish-brown and outlined in gilt. Light brown silk headbands. Turn-ins tooled with 3 light and 1 heavy gilt rule. Signed in gilt on turn-in of lower cover WILLIAM MACCANCE GREGYNOG PRESS BINDERY GEORGE FISHER. Price 12 guineas.

21 1932 **Caniadau gan W. J. Gruffydd**

[wood engraved shadow title in Tyrian red and black] Caniadau [double rule] gan: W. J. Gruffydd [Device No. 19] Gwasg Gregynnog MCMXXXII [double rule]

[v] vi–vii [1] 101 [3] pp. Published 9 July 1932. Perpetua type on stone-coloured Charles I Barcham Green hand-made paper. Title-page and Welsh text printed in Tyrian red and black. 4 wood engravings and shadow initial letters by Blair Hughes-Stanton. Compositor: John Hugh Jones; Pressman: Herbert John Hodgson; Completed 5 May 1932 'As at the time of printing Eric Gill had not finally approved the italic version of Perpetua for issue by the Monotype Corporation, those areas of the text intended to be in italic were printed in red' (colophon). 255 × 180 mm. 275 copies bound in paper printed from press roller inked with Tyrian red, with grey buckram spine. Titling in gilt on spine and front of spine strip. Unsigned. Price 1 guinea.

Special binding: 255 × 180 mm. First 25 copies bound in brown levant morocco, polished. Decorated with a panel of vertical gilt lines which connects the covers across the spine, echoed by half panels of the same at the outer edges which continue to the turn-ins. 5 raised bands outlined in blind. Gilt titling and imprint. Red silk headbands. Signed in gilt on turn-in of upper cover GREGYNOG PRESS BINDERY and in gilt on turn-in of lower cover BLAIR HUGHES-STANTON GEORGE FISHER. Price 6 guineas.

22 1932 **The singing caravan**

The singing caravan a Sufi tale by Robert Vansittart [quotation] [Arabic characters] [in brown and black Device No. 21] MCMXXXII The Gregynog Press Newtown Montgomeryshire Wales

[iv] v–vii [3] 142 [4] pp. Published 5 November 1932. Perpetua type on Barcham Green hand-made paper. Wood engraved frontispiece, tailpiece and initial letters by William MacCance. Printed in black and brown. Compositors: Idris and Idwal Jones; Pressman: Herbert John Hodgson; Completed 1 September 1932 (colophon). 280 × 172 mm. 225

copies bound in brown Welsh sheepskin with gilt spine titling and title blocked in gilt on upper cover, set in a pattern resembling musical notation. Unsigned. Price 2 guineas.

Special binding: 277×182 mm. First 25 copies bound in deep orange oasis, polished. Upper cover has a fore-edge flap. All-over intricate pattern of gilt lines which intersect and incorporate on each cover 4 areas onlaid in black oasis and outlined in gilt. The upper cover includes the author and title in gilt and the lower cover the imprint. Bands of 5 gilt lines of varying thickness at head and tail are carried across the spine and continue to the flap. Gilt spine titling combined with vertical and diagonal lines. Gold silk headbands. Turn-ins tooled with 5-rule fillet to intersect at the outer corners. Signed in gilt on turn-in of lower cover WILLIAM MACCANCE GREGYNOG PRESS BINDERY GEORGE FISHER. Price 10 guineas.

23 1933 **Erewhon**

Erewhon [double rule] by Samuel Butler The wood-engravings by Blair Hughes-Stanton [set at right angles to the title and running up from the foot] The Gregynog Press Montgomeryshire [rule] [Device No. 20] MCMXXXII [rule]

[6] 7–266 [2] pp. Published 11 February 1933. Baskerville type on Japanese vellum. 29 wood engraved head-pieces by Blair Hughes-Stanton. Running titles are ruled beneath. The music on page 51 was inscribed by Rosemary Ratcliffe and cut on wood by R. J. Beedham. Compositors: Richard Owen Jones and Idris Vaughan Jones; Pressman: Herbert John Hodgson; Completed 9 July 1932 (colophon). 230×150 mm. 275 copies bound in sage coloured Welsh sheepskin blocked on the spine in gilt with titling and device. Signed on paste-down of lower cover GREGYNOG PRESS BINDERY. Price 3 guineas.

Special binding: 230×148 mm. First 25 copies bound in brown levant morocco, polished. Tooled in gilt with an all-over design consisting of a horizontal band of 23 fine lines which starts on the lower cover and ends on the upper cover, the ends terminating in stirrup-like designs. Upper cover design incorporates title, author and imprint. 4 raised bands. Titling and imprint contained in gilt ellipses. Rust-brown silk headbands. Turn-ins tooled in gilt with 2 heavy and 1 fine gilt rule. Signed in gilt on turn-in of lower cover GREGYNOG PRESS BINDERY BLAIR HUGHES-STANTON. Price 8 guineas.

24 1933 The revelation of Saint John the Divine

[in Tyrian red and black] The revelation of Saint John the Divine [Device No. 22] The Gregynog Press MCMXXXII

[60] pp. Published 5 April 1933. Bembo type and Perpetua titling on Japanese vellum. Text printed in Tyrian red and black. The first 3 words of the title on the title-page are wood engraved. 41 wood engravings, title-page and book design by Blair Hughes-Stanton. Compositor: John Hugh Jones; Pressman: Herbert John Hodgson; Edition taken from the text of Dr. Scrivener's Paragraph Bible by consent of Cambridge University Press; Completed 30 November 1932 (colophon). 344×205 mm. 232 copies bound in red Hermitage calf, blocked in blind on upper cover with title and Device No. 22. Blind spine titling. Top edge sprinkled. Unsigned. Price 6 guineas.

Special binding: 342×200 mm. 18 copies (Nos. 1–15, 102, 242 and 243) bound in cream large grain levant morocco, unpolished, with black and brown calf onlays. The upper cover decorated with a crushed panel, outlined with a heavy gilt rule, which encloses a large onlaid black cross, outlined and shaded in gilt, bearing THE REVELATION in gilt. An adjoining brown onlay, outlined and shaded in blind, is tooled in blind with the remainder of the title. On the lower cover similar decoration on a smaller scale incorporates the gilt imprint. 5 raised bands outlined in blind. Spine titling and GG in gilt. Rust-brown silk headbands. Turn-ins decorated with 3 gilt rules of varying width. Signed in gilt on turn-in of lower cover BLAIR HUGHES-STANTON GREGYNOG PRESS BINDERY GEORGE FISHER. Price 12 guineas.

25 1933 Clych atgof gan O. M. Edwards

[in sepia and black] Clych atgof [double rule] Penodau yn hanes fy addysg gan: Owen Edwards [quotation] [Device No. 23] [double rule] Gwasg Gregynog ger y Drenewydd Sir Drefaldwyn MCMXXXIII

viii [2] 95 [3] pp. Published 15 July 1933. Perpetua type on Batchelor's special making of GG watermarked hand-made paper. Title-page and text printed in brown and black. Frontispiece and 8 decorations engraved on wood by William MacCance. Compositor: Richard Owen Jones; Pressman: Herbert John Hodgson; Completed 31 March 1933 (colophon). 212×140 mm. 385 copies bound in Welsh sheepskin. Titling and imprint printed in brown ink on spine and front cover of case. Unsigned. Price 1 guinea.

Special binding: 212 × 140 mm. First 15 copies bound in brown levant morocco, polished. Covers tooled in gilt with 2 horizontal bands of gilt lines which terminate in lozenges. 4 raised bands. Spine panels 2 and 3 outlined in blind, 1 and 4 in gilt lines which continue from the covers. Titling and imprint in gilt. Panel 3 decorated with 5 vertical gilt lines which terminate in lozenges. Gold silk headbands. Turn-ins tooled in gilt with 3 lines terminating in lozenges. Signed in gilt on turn-in of lower cover GREGYNOG PRESS BINDERY GEORGE FISHER. Price 5 guineas.

26 1933 **Four poems by John Milton**

Four poems by John Milton L'Allegro Il Penseroso Arcades Lycidas [wood engraving] with wood-engravings by Blair Hughes-Stanton The Gregynog Press MCMXXXIII

[2] 33 [3] pp. Published 19 October 1933. Perpetua type on Japanese vellum. 11 wood engravings and book design by Blair Hughes-Stanton. The text edited by the Rev. H. C. Beeching. Compositors: John Hugh Jones and Idwal Jones; Pressman: Herbert John Hodgson; Completed 5 January 1933 (colophon). 257 × 167 mm. 235 copies bound in reddish Hermitage calf blocked in blind on upper cover with the figure of Euphrosyne, titling and Press device. Spine tooled with blind lines. Unsigned. Price 2 guineas.

Special binding: 255 × 167 mm. First 14 copies bound in dun-coloured levant morocco, unpolished. The upper cover decorated with a vertical crushed panel, outlined in gilt, which is invaded by three horizontal panels also outlined in gilt. The top one is filled with intersecting blind lines and the middle one with parallel blind lines. The lower cover is decorated with the same pattern in reverse. 5 raised bands. Spine decorated with panels which echo those on the covers. Titling and imprint in blind. Buff silk headbands. Turn-ins tooled with a heavy blind and a fine gilt rule. Signed in blind on turn-in of lower cover BLAIR HUGHES-STANTON GREGYNOG PRESS BINDERY GEORGE FISHER. Price 7 guineas. *Variant:* No. 15 bound in rose-pink levant morocco, polished. Decorated with an all-over design of perpendicular lines in blind terminating in onlaid lozenges, outlined in blind. This binding designed by George Fisher.

27 1933 **XXI Welsh gypsy folk-tales**

[wood engraved shadow title] XXI Welsh gypsy folk-tales collected by: John Sampson [wood engraving] with engravings on wood by: Agnes

Miller Parker [wood engraving] The Gregynog Press Newtown Montgomeryshire – MCMXXXIII

[vi] vii–xi [3] 108 [2] pp. Published 14 November 1933. Bembo type on Portal's hand-made paper. 8 wood engravings by Agnes Miller Parker. Text edited by Dora E. Yates. Compositor: Idris Jones; Pressman: Herbert John Hodgson; Completed 12 June 1933 (colophon). Device No. 23 in black follows colophon. 277 × 190 mm. 235 copies bound in mustard-yellow Welsh sheepskin with title, set in a pattern of rules, blocked in gilt on the upper cover. Gilt spine titling. Unsigned. Price 3 guineas.

Special binding: 276 × 190 mm. First 15 copies bound in mustard-yellow levant morocco, polished, with black and red onlays. Covers decorated with panels of gilt lines at head and tail. Between these a vertical arrangement of black and red onlaid lozenges, outlined in gilt and connected by gilt lines. 5 raised bands outlined in blind. First and sixth spine panels decorated with gilt vertical lines. Gilt titling. Red silk headbands. Turn-ins decorated with oblique gilt lines at head and tail. 3 onlaid black and red lozenges, outlined in gilt, down fore-edge, linked by gilt lines. Signed in gilt on turn-in of lower cover GREGYNOG PRESS BINDERY GEORGE FISHER. Price 8 guineas.

28 1933 **The lovers' song-book**

The lovers' song-book by W. H. Davies [in blue (variants in green) and black Device No. 9] The Gregynog Press MCMXXXIII

v [1] 30 [2] pp. Published 1 December 1933. Baskerville type on Japanese vellum. Text pages ruled in blue (variants in green). Wood engraved initial letters in blue (variants in green). Completed 11 November 1933 (colophon). 230 × 155 mm. 231 copies bound in blue cloth (variants in green) spine, titled in gilt, with matching sides of Cockerell marbled paper. Unsigned. Price 1 guinea.

Special binding: 230 × 152 mm. 19 copies (Nos. 1–15, 143, 202, 226 and 227) bound in light green levant morocco, polished. Tooled in gilt with 11 foliate sprays rising vertically from a double line of lozenges. 2 raised bands, outlined in blind. Tooled in gilt with title and foliate sprays rising from lozenges. Yellow silk headbands. Turn-ins tooled in gilt with a double line of lozenges along the tail and a single leaf spray rising up the fore-edge. Signed in gilt on turn-in of lower cover GREGYNOG PRESS BINDERY GEORGE FISHER. Price 6 guineas.

29 1934 The lamentations of Jeremiah

[wood engraved title-page in blue and black] The lamentations of Jeremiah The Gregynog Press MCMXXXIII

[30] pp. Published 20 July 1934. Baskerville italic on heavy Japanese vellum. Title-page, 5 full page engraved chapter openings, 16 other wood engravings and book design by Blair Hughes-Stanton. Headings and running titles printed in blue. Edition taken from the text of Dr. Scrivener's Paragraph Bible by consent of Cambridge University Press; Compositor: John Hugh Jones; Pressman: Herbert John Hodgson; Completed 10 August 1933 (colophon). 385 × 255 mm. Copies numbered 16–129 and 240–250 bound in nocturnal blue Hermitage calf, copies 130–239 in blue oasis. Title and device blocked in blind on upper cover. Top edge sprinkled. Unsigned. Price 5 guineas.

Special binding: 385 × 250 mm. First 15 copies bound in black levant morocco, unpolished. A large disc onlaid in mid-blue oasis adjoins an area of blind and heavy gilt lines. From the disc hangs an avenging sword, tooled in gilt and blind. The disc is connected by a fine gilt line to a line of 6 onlaid white squares, outlined in gilt, connected to each other by gilt lines of varying width. 6 irregularly placed raised bands outlined in blind. Title and imprint in gilt. Panel 4 tooled in gilt with a long sword. Blue silk headbands. Turn-ins decorated with 1 heavy and 1 light gilt rule. Signed in gilt on turn-in of lower cover BLAIR HUGHES-STANTON GREGYNOG PRESS BINDERY GEORGE FISHER. Price 12 guineas.

30 1934 Don Quixote

Don Quixote an introductory essay in psychology by Salvador de Madariaga Translated from the Spanish by the author and Constance H. M. de Madariaga [in blue Device No. 11] The Gregynog Press MCMXXXIV

xv [1] 136 [2] pp. Published 14 September 1934. Baskerville type on Batchelor hand-made paper. Opening initials, chapter and running heads printed in blue. Completed 24 April 1934 (colophon). 244 × 170 mm. 235 copies bound in blue-grey Ingres paper with linen spine. Printed spine label. Unsigned. Price 2 guineas.

Special binding: 242 × 162 mm. First 15 copies bound in light blue levant morocco, polished. Simple all-over design consisting of intersecting panels of blind and fine gilt lines, surrounded by a gilt rule frame. 5 raised bands. Title and imprint in gilt. Cream silk headbands. Turn-ins tooled with 1 blind and 1 gilt rule. Signed in gilt on turn-in of the lower cover GREGYNOG PRESS BINDERY GEORGE FISHER. Price 6 guineas.

31 1934 Anne Boleyn and other poems

[in red] Anne Boleyn and other poems [in green Device No. 25] [in red] The Gregynog Press Newtown Montgomeryshire MCMXXXIV

77 [3] pp. Published 29 November 1934. Bembo type on Kelmscott hand-made paper. Text printed in red and black. Initial letters designed by Graily Hewitt, printed in red and green. Device No. 26 in green and red follows colophon. 210 × 167 mm. 285 copies bound in brown oasis, blocked in gilt on upper cover with crowned falcon badge. 3 imitation spine bands outlined in blind. Spine titling and imprint in gilt. Top edge gilt, fore-edge and tail uncut. Unsigned. Price 1 guinea.

Special binding: 210 × 165 mm. First 15 copies bound in dark brown oasis, polished. Both covers blocked in gilt with central crowned falcon badge surrounded by double gilt rule frame with solid gilt corner ornaments. 4 slightly raised bands. Spine gilt panelled. Gilt titling and imprint combined with crown and rose in panel 1 and rose on stem in panel 5. Light brown silk headbands. Turn-ins decorated in gilt with double rule fillet and fleurons at outer corners. Signed in gilt on turn-in of lower cover GREGYNOG PRESS BINDERY. Binding designed by Loyd Haberly. Price 4 guineas.

32 1935 The star of Seville

The star of Seville: a drama in three acts and in verse attributed to Lope de Vega translated out of Spanish by Henry Thomas [wood engraved shield in black, red, blue and orange] Gregynog MCMXXXV

[iv] v–xii [4] 108 [4] pp. Published 15 May 1935. Bembo type on Batchelor special making of GG watermarked hand-made paper. Text printed in red and black, initial letters in green. Arms of Seville on title-page designed by W. Charles Thomas. On recto following the colophon, Device No. 27 in red and green. 242 × 165 mm. 160 copies bound in black oasis blocked on both covers with a design of Moorish character, composed of interlaced blind strapwork and gold stars. Spine decorated in blind and gilt with strapwork and stars, titling and device in gilt. Top edge gilt. Unsigned. Price 3 guineas.

Special binding: 244 × 165 mm. First 15 copies bound in purple levant morocco, unpolished. Covers decorated with an onlaid blue-stained vellum star, outlined in gilt, from which radiate diagonal bands of gilt lines across a crushed panel. A horizontal band of scarlet onlay, outlined in heavy gilt, forms a base. 2 raised bands at the head of the spine, outlined in blind. Title in gilt in top panel. Below the bands an onlaid blue-stained vellum star from which hangs a band of gilt and blind lines terminating in a horizontal blind line. Underneath is a horizontal band of scarlet onlay, outlined in gilt, with the imprint in gilt at the foot. Purple silk headbands. Turn-ins ruled in gilt, inside which is a scarlet onlaid band, outlined in gilt. Signed in gilt on scarlet onlay on turn-in of lower cover GREGYNOG PRESS BINDERY BLAIR HUGHES-STANTON GEORGE FISHER. Price 8 guineas.

33 1935 **Eros and Psyche**

Eros and Psyche [wood engraving] A poem in XII measures by Robert Bridges: with wood-cuts from designs by Edward Burne-Jones Gregynog MCMXXXV

[8] 141 [3] pp. Published 29 October 1935. Gregynog type on Batchelor hand-made paper. Designs prepared for engraving by Dorothy Hawksley, 5 of which were cut on wood by Loyd Haberly and the remaining 19 by R. John Beedham. Initial letters designed by Graily Hewitt and printed in green. Text printed in red and black. 280 × 212 mm. 285 copies bound in white pigskin, blocked in gilt on upper cover with a circular design which incorporates hand tooling. Gilt ties to imitation spine bands. Gilt spine titling. Silk headbands to tone. Top edge gilt, fore-edge and tail uncut. Unsigned. Price 5 guineas.

Special binding: 280 × 210 mm. First 15 copies bound in light green oasis, polished. Covers tooled in palladium with 3 vertical bands of floral decoration. 4 raised bands outlined in blind. Titling and imprint tooled in palladium. Green silk headbands. Top edge palladium, fore-edge and tail uncut. Turn-ins tooled in palladium with decoration similar to that on the covers. Signed in palladium on turn-in of lower cover GREGYNOG PRESS BINDERY GEORGE FISHER. Price 12 guineas.

34 1936 **Cyrupaedia**

Cyrupaedia: the institution and life of Cyrus, the first of that name, King of Persians. Eight books treating of noble education, of princely exercises, military discipline, warlike stratagems, preparations & expeditions: written in Greeke by the sage Xenophon. [in red] Translated out of Greeke into English, and conferred with the Latine and French translations by Philemon Holland of the City of Coventry doctor in physick. [crown in black] [in red] Dedicated to his most excellent Majesty. [in black] First printed by J. L. for Robert Allot, to be sold at the Sign of the Beare in Paul's Churchyard, 1632, and now fully reprinted from the 1632 edition at the Gregynog Press, Newtown, Mont. MDCCCCXXXVI

[ii] iii–xvii [1] 321 [3] pp. Published 18 September 1936. Poliphilus type on Batchelor hand-made paper with hanging notes in red and floriated wood engraved initials by Loyd Haberly, hand coloured in red and green, at the opening of each book. Text edited by F. W. Bateson. 302×205 mm. 135 copies bound in dark green oasis. The covers decorated with centre- and cornerpieces onlaid in red and light green, outlined in gilt and blocked with gilt decoration of Persian character. Gilt rule frame. 5 raised bands. Gilt panelled spine with gilt title and imprint. Green silk headbands. Top edge gilt, fore-edge and tail uncut. Signed in gilt on turn-in of lower cover GREGYNOG PRESS BINDERY. Price 6 guineas.

Special binding: 300×205 mm. First 15 copies bound in heavy grained dark brown levant morocco. Covers decorated with a large central circular design which consists of a pattern of onlaid interlaced strapwork in red, orange, two shades of light brown and green, outlined in gilt, combined with areas of gilt tooling and blind intaglio blocking. The whole is surrounded by 2 recessed panels of orange onlay, outlined in gilt, and divided by a panel of blind lines. The onlaid colours appear in a different order on upper and lower cover. 4 raised bands. Spine panels framed with bands of coloured onlay, outlined in gilt. Spine titling and imprint in gilt. Panels 1, 4 and 5 are decorated with concentric blind panel frames. Brown silk headbands. Turn-ins decorated with 7-rule fillet in gilt. Signed in gilt on turn-in of lower cover GREGYNOG PRESS BINDERY GEORGE FISHER. Design adapted by Fisher from an original design by Lady Cartwright of Aynho. Price 15 guineas.

35 1936 **The story of a red-deer**

[in red] The story of a red-deer [in black] by the Honourable J. W. Fortescue With decorations by Dorothy Burroughes [illustration in green] The Gregynog Press MCMXXXV

[8] 125 [3] pp. Published 9 November 1936. Cloister type on Batchelor hand-made paper. Text printed in red and black. 11 illustrations by Dorothy Burroughes, printed in colour, to form head-pieces. Pressman: Herbert John Hodgson; Completed 8 November (colophon). 250×190

mm. 235 copies bound in red cloth, blocked in gilt on upper cover with a stag. Gilt spine titling. Top edge gilt, fore-edge and tail uncut. Unsigned. Price £1. 5*s*.
Special binding: 248 × 187 mm. First 15 copies bound in reddish brown oasis, polished. Covers richly tooled in gilt with a diagonal band of two alternating fir tree tools. Above this are fine vertical gilt lines and below heavier ones. Surrounded by gilt rule frames and 3 heavy gilt lines at the foot. 1 raised band. Spine tooled in gilt. Above the band are fine vertical lines with heavier ones beneath. 3 heavy gilt lines at the foot. Titling and imprint in gilt. Rust-brown silk headbands. Turn-ins tooled in gilt with fine and broad lines. Signed in gilt on turn-in of lower cover GREGYNOG PRESS BINDERY GEORGE FISHER. Price 7 guineas.

36 1937 **Caelica**
[in red] Caelica [in black] by Fulke Greville, Lord Brook Edited by Una Ellis-Fermor The Gregynog Press MCMXXXVI
[ii] iii–xiii [1] 149 [1] pp. Published 20 April 1937. Perpetua type on Arnold & Foster unbleached hand-made paper. Initial letters in red. Completed 22 October 1936. (colophon) 230 × 152 mm. 210 copies bound in green patterned paper, designed by James Wardrop, with broad spine strip of dark green oasis tooled in gilt with title, imprint and a pattern of lines. Oasis corner tips to match. Signed on paste-down of lower cover GREGYNOG PRESS BINDERY. Binding design by James and Evelyn Wardrop. Price £1. 15*s*.
Special binding: 230 × 150 mm. First 15 copies bound in heavy grained slate-blue levant morocco, unpolished. Decorated with an all-over design consisting of 16 onlaid vellum stars, stained pink and outlined in blind and gilt, which are connected by gilt and blind lines. 7 raised bands, outlined in gilt and blind with lines which continue from the covers. 4 onlaid vellum stars, stained pink. Title and imprint in gilt. Blue silk headbands. Turn-ins decorated with 1 blind and 1 gilt rule. Signed in gilt on turn-in of lower cover BLAIR HUGHES-STANTON GREGYNOG PRESS BINDERY GEORGE FISHER. Price 8 guineas.

37 1937 **The history of Saint Louis**
The history of Saint Louis by John, Lord of Joinville Seneschal of Champagne translated from the French text edited by Natalis de Wailly by Joan Evans The Gregynog Press MCMXXXVII

[6] 157 [11] pp. Published 8 December 1937. Poliphilus type on Arnold & Foster hand-made paper. Openings and initial letters, printed in red and pale blue, designed by Alfred J. Fairbank and cut on wood by R. John Beedham. 17 hand-coloured armorial shields, proved by A. van der Put and engraved on wood by Reynolds Stone. 2 maps, drawn by Berthold Wolpe and lettered by Alfred J. Fairbank. Completed 30 September 1937 (colophon). 340×235 mm. 183 copies bound in brown oasis blocked in gilt on upper cover with the arms of St. Louis. 5 raised bands. Title and imprint in gilt. Red silk headbands. Top edge gilt, fore-edge and tail uncut. Signed on paste-down of lower cover GREGYNOG PRESS BINDERY. Price 6 guineas.

Special binding: 340×235 mm. First 15 copies bound in blue levant morocco, polished. Tooled in gilt with a diaper of fleur-de-lys set in a trellis of diagonal lines. Gilt rule frame. 5 slightly raised bands. Panels gilt framed. Gilt titling and imprint. Blue silk headbands. Turn-ins tooled in gilt with double rule fillet and fleur-de-lys at outer corners. Signed in gilt on turn-in of lower cover GREGYNOG PRESS BINDERY. Binding designed by James Wardrop. Price 14 guineas. *Variants:* Nos. 190 and 191 bound in cream levant morocco tooled in gilt with the arms of St. Louis on a blue onlay on upper cover and gilt spine titling.

38 1938 **The lovers of Teruel**

The lovers of Teruel a drama in four acts in prose and verse by Juan Eugenio Hartzenbusch Translated from the Spanish by Henry Thomas The Gregynog Press 1938

[iv] v–x [2] 112 [2] pp. Published 2 November 1938. Bembo type on Batchelor hand-made paper. 5 initial letters designed by Alfred J. Fairbank, printed in red. Completed 14 September 1938 (colophon). 242×170 mm. 155 copies bound in red niger goatskin. Covers decorated with a blind stamped Moorish design of interlaced strapwork. 3 imitation spine bands outlined in blind. Gilt spine tooling. Top edge gilt, fore-edge and tail uncut. Signed on paste-down of lower cover GREGYNOG PRESS BINDERY. Price 2 guineas.

Special binding: 242×167 mm. First 20 copies bound in large grain sage-green levant morocco, polished, with scarlet and black oasis onlays. Covers decorated with blind lines which form concentric squares and a large linear scarlet onlay, outlined in gilt and bordered by black onlay. 7 raised bands in groups of 2–3–2 with small black onlay between the top and bottom pair. Scarlet onlays, outlined in gilt, in panels 2 and 3. Title and imprint in gilt. Scarlet silk headbands. Turn-ins decorated with 1 blind and 1 gilt rule. Signed in gilt on turn-in of lower cover GREGYNOG PRESS BINDERY GEORGE FISHER BLAIR HUGHES-STANTON. Price 9 guineas.

39 1938 The praise and happinesse of the countrie-life

The praise and happinesse of the countrie-life Written originally in Spanish by Don Antonio de Guevara Put into English by H. Vaughan, Silurist [rule] [quotation from Virgil's Georgics] [rule] [wood engraving] Reprinted from the edition of 1651, with an introduction by Henry Thomas, and wood engravings by Reynolds Stone At the Gregynog Press Newtown 1938

xv [1] 37 [3] pp. Published 29 November 1938. Bell type on Arnold & Foster hand-made paper. 7 wood engraved head- and tail-pieces by Reynolds Stone. Completed November 1938 (colophon). 178 × 118 mm. 380 copies bound in olive green paper with wood engraved title label on upper cover printed in red and black. Red oasis spine with gilt titling and blind floral fillet border. Corner pieces of red oasis. Signed on paste-down of lower cover GREGYNOG PRESS BINDERY. Natural silurian paper dustwrapper with red spine titling and red wood engraved circular title panel on front. Price ½ guinea.

Special binding: 179 × 118 mm. First 20 copies bound in orange oasis, polished. Both covers blocked in blind with oriental intaglio stamp. 1 heavy and 1 light blind frame which continues over both covers. Spine panelled in blind with blind titling. Orange silk headbands. Turn-ins tooled with 1 heavy and 1 light rule frame. Signed in blind on turn-in of lower cover GREGYNOG PRESS BINDERY GEORGE FISHER. Price 2 guineas.

40 1939 Shaw gives himself away

Shaw gives himself away an autobiographical miscellany The Gregynog Press 1939

[vi] vii–xi [1] 188 [2] pp. Published 6 November 1939. Baskerville type on Arnold & Foster green tinted hand-made paper. Wood engraved frontispiece portrait by John Farleigh. Completed 18 August 1939 (colophon). 247 × 170 mm. 275 copies bound in dark green oasis decorated with onlaid abstract design in orange, based on the author's initials.

3 symbols on upper cover and 1 on lower cover. Horizontal band of orange onlay at foot of both covers. The spine has one raised band beneath which is a block of orange onlay. Titled in orange. Signed on paste-down of lower cover GREGYNOG PRESS BINDERY PAUL NASH. Price 3 guineas. *Special binding:* 247 × 167 mm. 21 copies bound in very dark green oasis. Richly tooled in gilt with alternating fine and thick vertical lines, 3 of which are broken by a pattern of lines forming a lozenge. The whole surrounded by 1 thick and 1 fine line which form a frame for both covers. 2 raised bands. Spine bordered with 2 fine gilt lines. Gilt titling and imprint. Panels 1 and 3 decorated with lines forming a lozenge. Bright green silk headbands. Turn-ins tooled with fine gilt rule. Signed in gilt on turn-in of lower cover GREGYNOG PRESS BINDERY GEORGE FISHER. Price 9 guineas. *Variants:* 4 copies bound in very dark green levant morocco with a similar design except that the outer frames are enriched with 2 bands of light green onlay and a very fine onlay of red. Also designed by George Fisher.

41 1940 **Gweledigaetheu y bardd cwsc**

Gweledigaetheu y bardd cwsc: visions of the sleeping bard by Ellis Wynne translated by T. Gwynn Jones

[xi] [2] 2–213 [3] pp. Published ? Thomas Jones's bibliography cites 9 March 1940 which is clearly an error. Bembo type on Barcham Green hand-made paper. Bilingual text, Welsh on verso and English on recto. Frontispiece from a wood engraving by Blair Hughes-Stanton. The Welsh text was prepared by J. E. Caerwyn Williams. Pressman: Herbert John Hodgson. Completed 12 March 1940 (colophon). 277 × 200 mm. 155 copies bound in decorated cloth which is plum coloured with red and powdered gold design. Red oasis spine tooled in gilt with bilingual titling, imprint and 7 vertical rules. Signed on paste-down of lower cover GREGYNOG PRESS BINDERY. Price £1 5*s*.

Special binding: 275 × 200 mm. First 20 copies bound in large-grain purple oasis, polished. Tooled in gilt with a pattern composed of 5 groups of upward-pointing arrows surrounded by 1 light and 1 heavy gilt rule frame. Light frames pierced at the head by downward-pointing lines which converge. 2 raised bands. Gilt ruled panels. Title and imprint in gilt. Purple silk headbands. Turn-ins tooled with heavy gilt rule. Signed in gilt on turn-in of lower cover GREGYNOG PRESS BINDERY GEORGE FISHER. Price 9 guineas.

42 1940 Lyrics and unfinished poems

Lyrics and unfinished poems by Lascelles Abercrombie The Gregynog Press MCMXL

[iv] v–xii 82 [2] pp. Published 10 August 1940. Romulus type on Barcham Green hand-made paper. Introduction by Wilfrid Gibson. Pressman: Herbert John Hodgson; Completed 28 June 1940 (colophon). 285 × 190 mm. 155 copies bound in green and grey Cockerell marbled paper with small leather corner tips and dark green oasis spine which is lettered in gilt. Decorated on upper and lower surfaces of spine strip with a gilt rule terminating in a lozenge. Signed on paste-down of lower cover GREGYNOG PRESS BINDERY. Price £1. 5*s*.

Special binding: 285 × 192 mm. First 20 copies bound in large-grain grey-blue oasis, polished. Covers tooled in gilt with a pattern of vertical lines of varying thickness. 2 raised bands. Spine titling in gilt below which are gilt vertical lines of varying thickness. Blue silk headbands. Turn-ins ruled in gilt with lines which diminish in thickness from head to tail. Signed in gilt on turn-in of lower cover GREGYNOG PRESS BINDERY GEORGE FISHER. Price 10 guineas.

Books Printed for Private Circulation

43 1929 Elphin Lloyd Jones

[wood engraved portrait] Elphin Lloyd Jones Arnold House, Saint John's Wood Hildersham House, Broadstairs The Hall, Hampstead

[4] 5–8 [8] 9–16 pp. Baskerville type on Japanese vellum. 3 wood engravings by R. A. Maynard. Includes a facsimile of a school essay on 8 unnumbered pages. Privately printed at the Gregynog Press, March 1929 (colophon). 231 × 170 mm. Bound in blue-grey Ingres paper boards. Decorated on upper board with circular blue device consisting of the initials ELJ entwined in interlaced strapwork connected to an encircling band bearing the legend BARRY 22 JANUARY 1916 HAMPSTEAD 21 DECEMBER 1928. Unsigned.

44 1933 A theme with variations

A theme with variations by Thomas Jones, C.H., The Gregynog Press MCMXXXIII

221 [3] pp. Perpetua type on Batchelor hand-made paper. Completed

14 December 1933 (colophon). Device No. 11 follows the colophon. 215 × 140 mm. 175 copies bound in red cloth. Blocked on upper cover with TJ monogram in gilt. Gilt spine titling. Unsigned.

45 1938 **John Davies**

John Davies born 5th May 1882. Died 5th December 1937

29 [1] pp. Baskerville type on Cowan Extra Strong machine-made paper. Printed at the Gregynog Press for private circulation (colophon). Undated but printed in 1938. 227 × 170 mm. 150 copies bound in blue Ingres paper boards. Blue JD monogram within double rule frame printed on upper cover. Bound June 1938 by R. A. Jones. Unsigned.

Engraving by Horace Bray from
The Misfortunes of Elphin.

WHERE SHALL WISDOM BE FOUND?

AND WHERE IS THE PLACE OF UNDERSTANDING?

WISDOM IS FAIRER THAN THE SUN AND IS ABOVE ALL THE ORDER OF THE STARS. WISDOM IS RADIANT AND FADETH NOT AWAY. WISDOM EXALTETH HER SONS, AND TAKETH HOLD OF THEM THAT SEEK HER.

HE THAT LOVETH HER, LOVETH LIFE, AND HE THAT SEEKETH HER EARLY SHALL BE FILLED WITH GLADNESS. HE THAT GIVETH EAR UNTO HER SHALL DWELL SECURELY, FOR THERE IS IN WISDOM A SPIRIT OF QUICK UNDERSTANDING, ALONE IN KIND, MANIFOLD IN EXPRESSION, LOVING TOWARD ALL MEN, STEADFAST, SURE AND FREE FROM CARE. AT FIRST SHE WILL WALK WITH HIM IN STRANGE WAYS, AND WILL BRING FEAR AND DREAD UPON HIM AND TORMENT HIM WITH HER DISCIPLINE, UNTIL SHE MAY TRUST HIS SOUL AND TRY HIM BY HER LAWS. THEN WILL SHE SHOW HIM THE STRAIGHT WAY AND WILL GLADDEN HIM AND REVEAL TO HIM HER SECRETS. IF HE TRUST HER, HE SHALL BEQUEATH HER, AND HIS GENERATION SHALL HAVE HER IN POSSESSION.

WHENCE THEN COMETH WISDOM? AND WHERE IS THE PLACE OF UNDERSTANDING?

GOD ALONE UNDERSTANDETH THE WAY THEREOF AND HE KNOWETH THE PLACE THEREOF. FOR HE LOOKETH TO THE ENDS OF THE EARTH AND SEETH UNDER THE WHOLE HEAVEN. HE CREATED IT, YEA, AND SEARCHED IT OUT. AND UNTO MAN HE SAID, BEHOLD, THE LOVE OF THE LORD, THAT IS WISDOM: AND TO DEPART FROM EVIL IS UNDERSTANDING.

Insert from Christmas Card, 1938, printed in blue and black. (*E215.*)
(*Side margins much reduced; the others increased.*)

Handlist of the Ephemera

ENTRIES ARE ARRANGED CHRONOLOGICALLY ACCORDING to printing dates where these are preserved in the Press records. Where dates are unavailable, entries are arranged chronologically to coincide with events with which they are connected. Entries which cannot be dated, even approximately, are placed at the end of the sequence. Capitalisation has been standardised and kept to a minimum. The actual punctuation is preserved. Since many of the items are in paper covers which form an integral part of the piece described, the covers are included in the unnumbered pagination; measurements are of the sheets inside the folders. Items consisting of more than two leaves are normally sewn with coloured silk. This is not described as the colours frequently vary from copy to copy.

1922

E1 Christmas Card

Let him who reads, ponder; / And pondering, remember; / And remembering, act; / And acting, persevere. Printed at the Gregynog Press. Christmas. 1922.
[8] pp. 237 × 155 mm. Kennerley type on grey silurian and Japanese vellum paper. Grey silurian covers enclose a four-page insert, on the second recto of which appears a wood engraving, measuring 102 × 128 mm, signed RAM, depicting Gregynog Hall, beneath which is a Welsh translation of the verse appearing on the front cover and 'Gregynog. Nadolig. 1922.'

E2 Concert Programme

Songs & carols Gregynog. December 28th. 1922. Printed at the Gregynog Press.
8pp. 220 × 143 mm. Kennerley type on Abbey Mills machine-made paper. The full points in the title and imprint are retained, indicating Maynard's lack of typographical knowledge at this early date.

1923

E3 Journal

Gregynog[:] an occasional journal privately issued from the Gregynog Press Montgomeryshire Number one. January 1923
[4] pp. 220 × 140 mm. Kennerley type on hand-made paper. Covers only. No journal traced. No further issues.

E4 Exhibition Note

A note on the Gregynog Press [Device No. 1]
Single leaf. 280 × 215 mm. Kennerley type on hand-made paper. 20 lines ending, 'The first book, of which a few pages are shown in this case, is now being set up . . .'

E5 The League of Nations Union. Welsh Council. World Wireless Message
The children of Wales and world peace The League of Nations Union Welsh Council 6 Cathedral Road, Cardiff *Colophon*: Printed by R. A. Maynard at the Gregynog Press Montgomeryshire May 1923.
[4] pp. 190 × 140 mm. Kennerley type on machine-made paper. On the inner verso, a note concerning the origins of the message and on the inner recto, the message itself, printed in red and black. 2,000 copies printed.
Another version of the above, printed on a single leaf measuring 190 × 270 mm.

E6 Poster
Montgomeryshire Musical Festival [wood engraving in vermilion and black] Bach Newtown 17 May 1923 Designed and printed at the Gregynog Press Montgomeryshire April 1923
Single leaf. 510 × 380 mm. Caslon wood face and Kennerley type on machine-made paper. The wood engraving is a portrait of J. S. Bach, signed R. A. Maynard.

E7 Poster
Aberystwyth Musical Festival [wood engraving in red and black] Bach June 22, 23, 24, & 25, 1923 University Hall Designed and printed at the Gregynog Press Montgomeryshire May 1923
Single leaf. 510 × 380 mm. Caslon wood face and Kennerley type on machine-made paper. The wood engraving is a portrait of J. S. Bach, signed R. A. Maynard.

E8 Concert Programme
The National Council of Music Performance of Bach's St. Matthew Passion at Gregynog July 14th 1923 [Device No. 2] *Colophon*: Printed by R. A. Maynard at the Gregynog Press Montgomeryshire July 1923
[4] pp. 282 × 215 mm. Kennerley type on Abbey Mills machine-made paper.

E9 Concert Programme
The National Council of Music Programme of concert at Gregynog July 16th 1923 [Device No. 2] *Colophon*: Printed by R. A. Maynard at the Gregynog Press Montgomeryshire July 1923
[4] pp. 285 × 220 mm. Kennerley type on Abbey Mills machine-made paper.

E10 Poem
[Wood engraving] Rhyddid [by] J. Ceiriog Hughes Printed at the Gregynog Press June 1923
Single leaf. 382 × 201 mm. Kennerley type on buff coloured hand-made paper. The wood engraving by Maynard, measuring 103 × 123 mm, printed in blue and black, depicts a mountain and lake scene. The poem consists of two stanzas in Welsh.

E11 Exhibition Catalogue
Catalogue of an exhibition of paintings, sculpture, etc. held at Gregynog, August 1923 *Colophon*: The Gregynog Press August 1923
[4] pp. 285 × 220 mm. Kennerley type on Abbey Mills machine-made paper. Colophon preceded by Device No. 1.

E12 Pamphlet
To the Council and Senate of the University of Birmingham from the Geographical Association *Colophon*: Printed by R. A. Maynard at the Gregynog Press Montgomeryshire November 1923
[4] pp. 210 × 151 mm. Kennerley type on Abbey Mills machine-made paper. A plea for the University of Birmingham to set up a department of geography, printed at the instigation of J. H. Fleure, Hon. Secretary of the Geographical Association and Professor of Geography and Anthropology at University College Aberystwyth. 160 copies printed.

E13 Christmas Card
Christmas MCMXXIII [ornament] *Colophon*: The Gregynog Press December 1923
[8] pp. 150 × 125 mm. Kennerley type on blue-grey silurian paper. On first verso, 2 verses. On facing recto, a wood engraving depicting Plâs Dinam, tipped in, beneath which is the message, 'With greetings and best wishes for Christmas & the New Year from Mrs. Edward Davies and the Misses Davies Plas Dinam MCMXXIII' 120 copies printed.

1924

E14 Poem
Written on walking in the woods of Gregynog in Montgomeryshire [Wood engraving] *Colophon*: First published 1794, in *Poems on Several Occasions* by Mrs. Darwall (formerly Miss Whateley) & now reprinted at the Gregynog Press, June 1924
Single leaf. 450 × 182 mm. Kennerley type on hand-made paper. The wood engraving, measuring 55 × 75 mm, depicts a lawn and woods at Gregynog. The poem consists of 8 stanzas.

E15 Concert Programme
The National Council of Music Programme of concert at Gregynog July 14th 1924 [Device No. 2] *Colophon*: Printed by R. A. Maynard at the Gregynog Press Montgomeryshire July 1924.
[4] pp. 282 × 223 mm. Kennerley type on Abbey Mills machine-made paper.

E16 Christmas Card
[Set in wood engraving] Christmas 1924 *Colophon*: The Gregynog Press December 1924
[8] pp. 155 × 110 mm. Kennerley type on silurian and hand-made paper. Title on blue-grey silurian paper cover, enclosing 4-page insert, on the first recto of which is printed, 'With Greetings and Best Wishes for Christmas and the New Year from Mrs. Edward Davies and the Misses Davies Gregynog 1924'. On the next verso and recto are printed five verses of a Christmas carol and a circular wood engraving depicting a shepherd.

E17 Concert Programme
A programme of Christmas music Gregynog Dec. 30th 1924 [Device No. 2] An exhibition of drawings and work carried out by the Gregynog Press during 1924

will be on view in the Library *Colophon*: Printed by R. A. Maynard at the Gregynog Press Montgomeryshire Dec. 1924
[6] pp. 290 × 222 mm. Kennerley type on Abbey Mills machine-made paper. The two-page insert, giving programme details, measures 278 × 217 mm.

1925

E18 Description
Gwasg Gregynog
Single leaf. 152 × 204 mm. Garamond type on card. A descriptive piece about the Press, consisting of 19 lines in Welsh.

E19 Subscribers' Form
Subscribers' form To Mr. R. A. Maynard, the Gregynog Press, Newtown, Montgomeryshire:—
Single leaf. 210 × 150 mm. Garamond type on machine-made paper. The form invites subscriptions for *Poems by George Herbert*, *Poems by Henry Vaughan* and *Caneuon Ceiriog*.

E20 Concert Programme
[Within a border of type ornaments] A programme of music to be sung by the Aberystwyth Small Choir at Gregynog on Saturday 31st. October 1925 [Device No 3] *Colophon*: Printed by R. A. Maynard at the Gregynog Press Montgomeryshire Oct. 1925.
[14] pp. 260 × 170 mm. Garamond type on Abbey Mills machine-made paper. There are variant forms of the cover which are either the same as the text paper or blue-grey silurian paper.

E21 Prospectus
The Gregynog Press [Device No. 5 white on red] December 1925
[12] pp. 265 × 200 mm. Garamond type on hand-made paper. Includes an account of the Press, a description of the first three books published and five specimen pages, printed in red and black.

E22 Christmas Card
[Device No. 4 in red] Christmas 1925 *Colophon*: The Gregynog Press
[8] pp. 188 × 107 mm. Garamond type on hand-made paper. Printed on one side of the sheet only. Third recto bears the message, 'To bring greetings and best wishes for Nineteen Twenty-six from Gregynog' and four wood engraved holly leaves, surrounding a dot, printed in black and red.

1926

E23 Prospectus
Detholiad o ganiadau gan T. Gwynn Jones Cynnwys yr argraffiad hwn bum can copi o bapur Van Gelder. Y mae naw addurniad y testun wedi eu torri ar bren gan Robert Ashwin Maynard a Horace Walter Bray. A limited edition of five hundred copies on Van Gelder paper. With nine woodcut decorations by Robert Ashwin Maynard and Horace Walter Bray. Gwasg Gregynog MCMXXVI Wyn-Edwards & Morris, 61, Renshaw Street, Liverpool.

[4] pp. 265 × 200 mm. Baskerville type on hand-made paper. Includes a bilingual description of the edition and a specimen page.

E24 Prospectus

Chosen essays by Edward Thomas [wood engraved portrait] The Gregynog Press MCMXXVI

[4] pp. 255 × 193 mm. Garamond type on hand-made paper. Includes a note on the work by Edward Garnett, a description of the edition and a specimen page.

E25 Conference Programme

League of Nations Union Welsh National Council Advisory Education Committee [4 ornaments] Gregynog October 15–18 1926 [Device No. 6] *Colophon*: Printed by R. A. Maynard at the Gregynog Press October 1926.

[12] pp. 202 × 125 mm. Poliphilus type on blue silurian paper covers and Abbey Mills machine-made paper. Includes a wood engraving of Gregynog Hall and 10 lines by Francis Thompson, a list of delegates and agenda. Printed on one side of sheet only.

E26 Concert Programme

A programme of vocal and chamber music Gregynog October 30th 1926 [Device No. 2] *Colophon*: Printed by R. A. Maynard at the Gregynog Press October 1926

[4] pp. 287 × 220 mm. Garamond type on Abbey Mills machine-made paper. The inside of the programme is set within a border of type ornaments.

E27 Christmas Card

To bring greetings and best wishes for Nineteen Twenty Seven from Gregynog *Colophon*: Being verses 6–21 from Chapter II of the Gospel according to Saint Luke. Reprinted at Gregynog from the Bishop Morgan Bible, 1588.

[6] pp. 255 × 210 mm. Poliphilus type on blue silurian and hand-made paper. The text, in Welsh, is printed in red and black on one side of a single leaf, measuring 255 × 210 mm, loosely inserted in blue silurian covers. The text opens with a large wood engraved decorated initial A which is white on red and occupies nine lines of type.

1927

E28 Loyal Address

Amgueddfa Genedlaethol Cymru National Museum of Wales [ornament] Formal opening by His Majesty King George V accompanied by Her Majesty Queen Mary on the 21st April, 1927 [ornament] The loyal address

[12] pp. 247 × 185 mm. Poliphilus and Blado italic types on hand-made paper. Bilingual text printed in red, blue and black with wood engraved opening and wood engraved initials in red and pale blue. Bound in blue levant morocco with title lettered in gilt in a panel on the upper cover surrounded by the words AMGUEDDFA GENEDLAETHOL CYMRU – NATIONAL MUSEUM OF WALES – AMGUEDDFA GENEDLAETHOL CYMRU – THE LOYAL ADDRESS in double gilt rule frame. The remainder of the upper and lower covers tooled in gilt with a semée of small foliate tool, alternating with square dots. Outer double rule frame in gilt.

E29 Exhibition Catalogue
Montgomeryshire Hospitals Appeal a hand list of the pictures and sculpture at Gregynog June – August 1927 *Colophon*: Printed by R. A. Maynard at the Gregynog Press June 1927.
[8] pp. 220 × 145 mm. Baskerville type on Abbey Mills machine-made paper.

E30 Prospectus
Selected poems of Edward Thomas [wood engraved ornament in blue] With an introduction by Edward Garnett [in blue] The Gregynog Press MCMXXVII
[4] pp. 230 × 150 mm. Garamond type on Japanese vellum. Inner pages ruled in blue. Text, in black with red wood engraved initial, consists of a note about the work, a specimen page and description of the edition.

E31 Conference Programme
[Within a border of type ornaments] The National Council of Music Conference at Gregynog July 8th–12th 1927 [Device No. 3] *Colophon*: Printed by R. A. Maynard at the Gregynog Press July 1927.
[4] pp. 260 × 171 mm. Garamond type on Abbey Mills machine-made paper. The inner pages, headed by a border of type ornaments, give details of musical events taking place during the conference.

E32 Concert Programme
[Within a border of type ornaments] A programme of vocal and chamber music Gregynog Monday 11th July 1927 [Device No. 3] *Colophon*: Printed by R. A. Maynard at the Gregynog Press July 1927.
[4] pp. 260 × 171 mm. Garamond type on Abbey Mills machine-made paper. The inner pages are headed by a border of type ornaments. Single leaf, loosely enclosed, measuring 255 × 166 mm of the libretti of Bach's *Wachet Auf* and Parry's *Jerusalem*.

E33 Prospectus
A list of editions printed at the Gregynog Press, with particulars of two books to be issued during the autumn of [outline wood engraved shadow numerals in red] 1927 . . .
[4] pp. 250 × 190 mm. Poliphilus type on hand-made paper. List printed in red and black.

E34 Subscribers' Form
Subscribers' form To Mr. R. A. Maynard, The Gregynog Press, Newtown, Montgomeryshire. Please send me: ...
Single leaf. 217 × 158 mm. Garamond type on machine-made paper. Invites subscriptions for *The Life of Saint David*, *Llyfr y Pregeth-wr*, *Poems by Henry Vaughan* and *Caneuon Ceiriog*.

E35 Christmas Card
[Within a border of type ornaments in blue] With greetings and best wishes for Christmas and the New Year from Mrs. Edward Davies and the Misses Davies Broneirion [in blue] 1927 Gregynog
Single leaf. 90 × 172 mm. Garamond type on blue card with linen finish.

E36 Christmas Card

Nineteen [wood engraved initial W in red, occupying four lines of type] ith greetings and Twenty best wishes for Seven Christmas and the New Year from Mrs. Edward Davies and the Misses Davies

[6] pp. 158 × 230 mm. Poliphilus type on Japanese vellum. The covers enclose a single leaf, measuring 155 × 220 mm, of Ben Jonson's *A hymn on the nativity of my saviour*, printed in black with red initials in a red double rule frame.

1928

E37 Prospectus

A list of editions printed at the Gregynog Press, with particulars of two books completed during the Spring of [outline wood engraved numerals in red] 1928 . . .

[4] pp. 253 × 190. Poliphilus type on Abbey Mills machine-made paper. Printed in red and black. Lists the first ten titles and announces the forthcoming publication of *Penillion Omar Khayyâm*, *Psalmau Dafydd*, *The Misfortunes of Elphin* and *Selected Poems* by W. H. Davies.

E38 Prospectus

The Gregynog Press, Newtown, Montgomeryshire Notice to permanent subscribers The autobiography of Edward Lord Herbert of Cherbury With an introduction by C. H. Herford and 9 wood engravings by Horace W. Bray The above volume, which will be ready for issue to subscribers early in June . . .

Single leaf. 210 × 203 mm. Poliphilus type on Japanese vellum.

E39 Invitation

Gregynog, Newtown, Montgomeryshire Telephone Newtown 66 The League of Nations Union Miss Henni Forchammer, the representative of Denmark at the assembly of the League at Geneva will speak in the Music Room, Gregynog, on 'Women and the League' on Saturday June 2nd 1928 at 3 o'clock You are cordially invited. R.S.V.P. The Secretary, Gregynog

Single leaf. 209 × 150 mm. Baskerville type on Japanese vellum.

E40 Prospectus

[Within a border of type ornaments] The Gregynog Press begs to announce that the following books, in limited editions, will be ready before the end of November, 1928 . . .

Single leaf. 139 × 183 mm. Baskerville type on hand-made paper. Books announced are:— *Penillion Omar Khayyâm*, *Selected Poems of W. H. Davies* and *The Misfortunes of Elphin.*

E41 Appeal

Distress in the mining areas of Wales and Monmouthshire Privately printed for the Welsh School of Social Service Hon. Secretary Edward Jones 22 High Street Newtown Montgomeryshire *Colophon*: Printed by R. A. Maynard at the Gregynog Press December 1928

[4] pp. 238 × 172 mm. Baskerville type on Abbey Mills machine-made paper. The title is printed in red and black.

E42 Christmas Card

To bring greetings and best wishes for Nineteen Twenty Nine from Gregynog Christmas 1928 1628–1928 Passages taken from John Bunyan's The Pilgrim's Progress

[6] pp. 257 × 210 mm. Poliphilus type on blue silurian and hand-made paper. The text is printed in red and black on a single leaf, loosely inserted in blue silurian paper covers, and opens with a large wood engraved decorated capital T, white on red, which occupies eight lines of type.

E43 Christmas Card

With greetings [wood engraved initial A in red, occupying four lines of type, enclosing the letters ND] best wishes for Christmas the New Year Mrs. Edward Davies Broneirion The Misses Davies Gregynog MCMXXVIII

Single leaf. 92 × 150 mm. Poliphilus type on hand-made paper.

A variant of the same card is printed on vellum.

1929

E44 Message

Gregynog Newtown Montgomeryshire The following message came from Dr. and Mrs. Tom Jones to the gathering at Gregynog Jan. 5th to 7th 1929 "Give our affectionate greetings to the group and should they wonder about us say we feel freshly dedicated to helpfulness of those whose sufferings and losses are more hard to bear than our own, with fewer to share them". Ever, T. J.

Single leaf. 202 × 138 mm. Baskerville type on Japanese vellum.

E45 Libretti

Jesu, priceless treasure Johann Sebastian Bach and Lord, it belongs not to my care Henry Walford Davies Gregynog April 26th. 1929

[4] pp. 223 × 142 mm. Baskerville type on Abbey Mills machine-made paper.

E46 Prospectus

The Gregynog Press A preliminary announcement May 1929

Single leaf. 210 × 140 mm. Baskerville type on Abbey Mills machine-made paper. Announces *Euripides*, *The stealing of the mare*, *Poems of Christina Rossetti* and *The essays and Last essays of Elia.*

E47 Certificate

[Within a wood engraved chevron border] Eisteddfod Genedlaethol Urdd Gobaith Cymru Fach Mehefin 1.9.2.9 Cor [wood engraving] wen Hyn sydd i arwyddo i o ennill clod ac anrhydedd yr eisteddfod yn am Gwasg y Gregynnog

Single leaf. 325 × 285 mm. Caslon type on hand-made paper. The border measures 170 × 185 mm. The wood engraving is a black outline shield containing a red dragon.

E48 Prospectus

19 September 29 [Device No. 10, white on red] An announcement of five books in active preparation at the Gregynog Press together with a list of books previously issued . . .

[4] pp. 283 × 217 mm. Garamond type on hand-made paper. Announces *Elia and The last essays of Elia*, *The plays of Euripides*, *Poems of Christina Rossetti*, *The stealing of the mare* and *Psalmau Dafydd* and lists the first thirteen books published. Includes two wood engravings from *The stealing of the mare.*

E49 Prospectus
The Gregynog Press announce [in red] Psalmau Dafydd [in black] Y cyfieithiad cyntaf i'r Gymraeg, gan William Morgan D.D., 1558, a than olygiaeth yr Athro Ifor Williams, M.A., Bangor.
[4] pp. 285 × 215 mm. Poliphilus type on Abbey Mills machine-made paper. The initial letter of Psalmau is a wood engraved letter, the tail of which occupies ten lines of type. The announcement includes a description of the edition in Welsh and English and a specimen page, printed in red and black.

E50 Acknowledgment
The Gregynog Press, Newtown, Montgomeryshire In accordance with your instructions received . . . we are reserving Nos. . . . of . . . The books will be sent to you as soon as the edition is ready for issue
Single leaf. 140 × 90 mm. Baskerville type on card. The specimen examined is signed by Dora Herbert Jones and dated November 1929.

E51 Christmas Card
May Christmas bring good cheer and the New Year all blessings from Mrs. Edward Davies, Broneirion and the Misses Davies, Gregynog 1929–1930
[4] pp. 160 × 115 mm. Poliphilus type on hand-made paper. The inside consists of a wood engraving and three lines by Herrick.

1930

E52 Concert Programme
[Within double red rules] The Saint Matthew Passion John Sebastian Bach [asterisk] given at Gregynog April 14th 1930 [Device No. 3] *Colophon*: Printed by R. A. Maynard at the Gregynog Press, Montgomeryshire, April, 1930
[4] pp. 290 × 220 mm. Baskerville type on Abbey Mills machine-made paper. The text is printed within double red rules.

E53 Libretto
The Saint Matthew Passion John Sebastian Bach 1685–1750 [asterisk] *Colophon*: Printed by R. A. Maynard at the Gregynog Press, Montgomeryshire, April, 1930
[2] 8 [2] pp. 283 × 219 mm. Baskerville type on stone cover paper and Abbey Mills machine-made paper. The text is printed within double red rules.

E54 Bookplate
[Circular wood engraving in red] This volume is one of a collection of books given to Coleg Harlech in memory of Elphin Lloyd Jones
Single leaf. 105 × 83 mm. The wood engraving consists of rope-work ornament with the initials ELJ intertwined, surrounded by the words, 'Barry 22 January 1916 . . . Hampstead 21 December 1928' which surmounts the legend. Gummed on the reverse side.

E55 The League of Nations Union. Welsh Council. World Wireless Message
The world wireless message of the children of Wales Neges ddiwifr ieuenctid Cymru i'r byd Printed at the Gregynog Press, June, MCMXXX
Single leaf. 295 × 380 mm. Poliphilus type on hand-made paper. Text in Welsh and English printed in two columns, divided by a wood engraved dragon, printed in red.

E56 The League of Nations Union. Welsh Council. World Wireless Message
[in red] The world wireless message of the children of Wales [in black] 'Goodwill Day', May 18th
Single leaf. 257 × 204 mm. Baskerville type on Japanese vellum. Item consists of a letter acknowledging replies received, from Gwilym Davies, Vice-president of the Council.

E57 Certificate
[within a wood engraved chevron border] Eisteddfod Genedlaethol [in red] Urdd Gobaith Cymru [in black] 19. Mai. 30 Caern [wood engraving] arvon Hyn sydd i arwyddo i o ennill clod ac anrhydedd yr eisteddfod yn am Gwasg y Gregynnog
Single leaf. 237 × 315 mm. Caslon type on hand-made paper. The border measures 170 × 185 mm. The wood engraving is a black outline shield containing a red dragon.

E58 Concert Programme
The Saint Matthew Passion (abridged) John Sebastian Bach 1685–1750 [Device No. 3] *Colophon*: Printed by R. A. Maynard at the Gregynog Press, Montgomeryshire, April, 1930
[2] 8 [2] pp. 287 × 224 mm. Baskerville type on double stone cover paper and Abbey Mills machine-made paper. Text pages ruled in red with double rule. Includes lists of artists who participated at performances given at Gregynog on 14 July 1923; 14 April 1930 and 12 July 1930, together with the libretto. A slip, sewn round text, notes, 'The Gregynog Small Choir not to be taken away'. Enclosed is a single programme leaf, measuring 280 × 210 mm, printed on one side only.

E59 Announcement
The Gregynog Press Newtown, Montgomeryshire The Misses Davies of Gregynog announce that on the retirement of Mr. R. A. Maynard and Mr. H. W. Bray from the Gregynog Press, they have secured the services of a group of four young artists . . .
Single leaf. 255 × 204 mm. Baskerville type on Japanese vellum. 21 lines announcing the appointment of Blair Hughes-Stanton, Gertrude Hermes, Agnes Miller Parker and William MacCance.

E60 Conference Programme
The Burge Memorial Trust International Conference Levavi [Device No. 12] oculos Gregynog 26–29 September 1930 *Colophon*: Printed by R. A. Maynard at the Gregynog Press September 1930
[16] pp. 202 × 123 mm. Baskerville type on double stone cover paper and Abbey Mills machine-made paper. Includes a wood engraving of Gregynog Hall, 10 lines by Francis Thompson, a list of members of the Trust and a list of guests.

E61 Order of Service

The Burge Memorial Trust International Conference 26 – 29 September 1930 [cross in red] An order of service to be used at Gregynog on Sunday 28 September 1930 [cross in red] *Colophon*: Printed by R. A. Maynard at the Gregynog Press September 1930

9 [3] pp. 287 × 188 mm. Baskerville type on double stone cover paper and Abbey Mills machine-made paper. Text printed in black and red. On last verso of text, 'Printed by R. A. Maynard at the Gregynog Press September 1930'. The first order of service to be printed at Gregynog. On lower cover, 'Levavi [Device No. 12] oculos'.

E62 Conference Programme

League of Nations Union Welsh National Council Advisory Education Committee [wood engraving under rule] Gregynog 14–17 November, 1930 *Colophon*: Gwasg.y.Gregynnog 1930

[12] pp. 205 × 123 mm. Garamond type on double light brown cover paper and Abbey Mills machine-made paper. Contains 7 lines by Robert Bridges, two wood engravings, a list of committee members, guests and agenda. On penultimate text verso, 'Planned by William MacCance. Engraved by Blair Hughes-Stanton, and carried out by the Staff of the Gregynog Press. November 1930'. Colophon preceded by Device No. 14 and the motto 'Levavi oculos'.

E63 Order of Service

League of Nations Union Welsh National Council Advisory Education Committee 14–17 November 1930 [cross] An order of service to be used at Gregynog on Sunday 16 November 1930 [cross] *Colophon*: Gwasg.y.Gregynnog 1930

[2] 11 [3] pp. 290 × 187 mm. Garamond type on double light brown cover paper and Abbey Mills machine-made paper. On penultimate text verso, 'Planned by William MacCance. Engraved by Blair Hughes-Stanton, and carried out by the Staff of the Gregynog Press. November 1930'. Colophon preceded by Device No. 14 and the motto 'Levavi oculos'.

E64 Christmas Card

[Underlined with double rule] Greetings and best wishes for Christmas and the New Year [wood engraving] from the Gregynog Press [underlined with double rule] MCMXXX

[8] pp. 254 × 162 mm. Bembo type on Japanese vellum. Printed on one side of the sheet only. The wood engraving depicts a dove with olive branch. The inside contains three further wood engravings and lines from Robert Bridges's *Testament of Beauty*. On lower cover, Device No. 15.

E65 Christmas Card

[Underlined with double rule] Greetings and best wishes for Christmas and the New Year [wood engraving] from the Misses Davies Gregynog [underlined with double rule] MCMXXX

[8] pp. 254 × 162 mm. Bembo type on Japanese vellum. Printed on one side of the sheet only. The wood engraving depicts a dove with olive branch. The inside

contains three further wood engravings and lines from Robert Bridges's *Testament of Beauty*. On lower cover, Device No. 15.

E66 Christmas Card

[Wood engraving under double red rule, then underlined with double red rule] Greetings 1930 [asterisk in red] 1931

[8] pp. 155 × 115 mm. Bembo type on Japanese vellum. Printed on one side of the sheet only. Inside: message printed in red and black with wood engraved initial B and holly leaf, 'Best wishes and * kind thoughts ** for Christmas and the New Year from Mrs. Edward Davies **** Broneirion The Misses Davies ***** Gregynog'. On lower cover, Device No. 15.

1931

E67 Invitation

The Gregynog Press [asterisk] We should like to invite you to an exhibition of fine books opening on 9th March 1931 in the Board Room of the Old Court House at the bookshop of Messrs. John & Edward Bumpus Ltd. of 350, Oxford Street, London. There will be a representative collection of books issued from this Press and a comprehensive display of the work of the private presses.

Single leaf. 99 × 133 mm. Caslon type on Japanese vellum.

E68 Subscription List

The Elphin Lloyd Jones Memorial Fund [wood engraving in red] The Gregynog Press MCMXXX

28 pp. 282 × 205 mm. Bembo type on hand-made paper. Bound in red buckram, blocked in gilt on the upper cover with the same decorative device on the title-page which consists of Celtic rope-work ornament with the initials ELJ intertwined, surrounded by the words, 'Barry 22 January 1916 . . . Hampstead 21 December 1928'. Text consists of a copy of the trust deed establishing the fund and a list of nearly 340 subscribers who gave a total of £3,067. 0*s*. 4*d*. Printing completed 31 March 1931.

E69 Invitation

Chamber and vocal music to which you are invited will be given at Gregynog at 6-30 p.m., on Monday, April 27th, 1931 R.S.V.P. to the Secretary, Gregynog, nr. Newtown

Single leaf. 88 × 114 mm. Garamond type on card which has rounded corners.

E70 Concert Programme

A concert of chamber and vocal music [ornament] Gregynog 27th April 1931 *Colophon*: The Gregynog Press 1931 [Device No. 16]

[8] pp. 277 × 185 mm. Poliphilus type on Abbey Mills machine-made paper. Printed on one side of the sheet only.

E71 The League of Nations Union. Welsh Council. World Wireless Message

The tenth annual wireless message of the youth of Wales Degfed neges radio flynyddol ieuenctid Cymru i'r byd Printed at the Gregynog Press

Single leaf. 310 × 415 mm. Poliphilus type on hand-made paper. Text in Welsh and

English printed in two columns, divided by a wood engraved dragon, printed in red. Each column is dated at the foot 'May 18, 1931'.

E72 The League of Nations Union. Welsh Council. World Wireless Message
[in green] The world wireless message of the youth of Wales [in black] 'Goodwill Day' May 18th
Single leaf. 256 × 205 mm. Baskerville type on hand-made paper. Item consists of a letter acknowledging replies received, from Gwilym Davies, Vice-president of the Council. At the foot in green, '1931 Welsh League of Nations Union 1932'.

E73 Certificate
[Within a wood engraved chevron border] Eisteddfod Genedlaethol Urdd Gobaith Cymru 19. Mai. 31 Aber [wood engraving] tawe Hyn sydd i arwyddo i o ennill clod ac anrhydedd yr eisteddfod yn am Gwasg Gregynog
Single leaf. 354 × 235 mm. Caslon type on hand-made paper. The border measures 170 × 185 mm. The wood engraving is a black outline shield containing a red dragon.

E74 Prospectus
Euripides [double rule] The following plays Hippolytus Electra The Bacchae The Trojan Women Medea Iphigenia in Taurus Rhesus Alcestis translated into English rhyming verse by Gilbert Murray, LL.D., D. Litt. with wood-engravings from the Greek vase paintings by Robert Ashwin Maynard and Horace Walter Bray . . .
[4] pp. 380 × 250 mm. Bembo type on hand-made paper. Contains a description of the edition and a specimen page.

E75 Prospectus
The Gregynog Press announces Milton's Comus . . .
Single leaf. 100 × 135 mm. Baskerville type on Japanese vellum. A description of the edition.

E76 Prospectus
[Wood engraved title] Comus a mask by John Milton with a frontispiece and the six characters in costume designed and engraved on wood by Blair Hughes-Stanton [wood engraving] The Gregynog Press MCMXXXI
[8] pp. 275 × 175 mm. Baskerville type on Japanese vellum. Contains four specimen pages and letters from H. Lawes to John, Lord Viscount Bracly and from Sir Henry Wotton to the author.

E77 Invitation
Mass in G. Minor R. Vaughan Williams to be given at Gregynog 11 July, 1931 at 8 p.m. The audience is requested to be seated at 7.50 p.m. Please bring this card with you
Single leaf. 88 × 140 mm. Baskerville type on card.

E78 Concert Programme
A concert of chamber and vocal music [ornament] Gregynog 11 July 1931

Colophon: The Gregynog Press 1931 [Device No. 16]
[8] pp. 275 × 185 mm. Poliphilus type on Abbey Mills machine-made paper. Printed on one side of the sheet only.

E79 Libretto
Mass in G. Minor for soli and double chorus [double rule] by R. Vaughan Williams [Device No. 17] *Colophon*: The Gregynog Press [Device No. 16]
[8] pp. 272 × 183 mm. Poliphilus type on grey silurian cover paper and hand-made paper. Covers printed in sepia. Sewn in wrap-round slip on which is printed, 'The Gregynog Choir. Not to be taken away'.

E80 Order of Service
National Council of Music twelfth annual conference 11–12 July 1931 [cross] An order of service to be used at Gregynog on Sunday 12 July 1931 [cross] *Colophon*: Gregynog Press 1931
[2] 8 [2] pp. 287 × 188 mm. Garamond type on double stone cover paper and Abbey Mills machine-made paper. On final text recto, 'Printed at the Gregynog Press. July, 1931'. Colophon preceded by Device No. 17.

E81 Book (unpublished)
The lovers' song book by W. H. Davies [wood engraving] With wood engravings by Blair Hughes-Stanton The Gregynog Press MCMXXXI
vi, 30 pp. 232 × 155 mm. Poliphilus type on Japanese vellum. Printed in black and reddish-brown. First edition of the book, rejected by the Press Board and not published.

E82 Invitation
Gregynog, Newtown, Montgomeryshire Telephone Newtown 66 You are invited to view an exhibition of work done in the distressed areas of South Wales, to be opened by Sir Percy E. Watkins, LL.D., Permanent Secretary, Welsh Department, Board of Education, at Gregynog on Wednesday, 12th August, 1931 at 3.0 p.m. Representatives from the various centres will be present and will give short accounts of the recent developments in craftwork and in social reconstruction in these areas. Articles may be purchased and orders given. R.S.V.P. The Secretary Gregynog
Single leaf. 210 × 138 mm. Baskerville type on Japanese vellum.

E83 Exhibition Opening Programme
Exhibition of work done in the distressed areas of South Wales, to be opened by Sir Percy Watkins, LL.D., Permanent Secretary, Welsh Department, Board of Education, at Gregynog on Wednesday, 12th August, 1931 at 3.0 p.m.
[4] pp. 190 × 138 mm. Baskerville type on Abbey Mills machine-made paper. Contains a list of centres exhibiting work and the words of the hymn *Jerusalem*.

E84 Appeal
Syr John Rhŷs 1840 . . 1915 *Colophon*: Gregynog Press 1931
[8] pp. 278 × 190 mm. Poliphilus type on dark brown cover paper and hand-made paper. An appeal for funds to erect a village hall in Ponterwyd, Cardiganshire, the

native village of Sir John Rhŷs, in his memory. Signed wood engraved frontispiece portrait by William MacCance. Texts in English and Welsh are separated by an unsigned wood engraving of Abercaero, Rhŷs's birthplace. Colophon preceded by Device No. 17.

E85 Order of Service

International Voluntary Service 5–6 September 1931 [cross] An order of service to be used at Gregynog on Sunday 6 September 1931 [cross] *Colophon*: Gregynog Press 1931

[2] 10 pp. 285 × 187 mm. Garamond type on blue cover paper and Abbey Mills machine-made paper. Colophon preceded by Device No. 17.

E86 Order of Service

League of Nations Union Welsh National Council Advisory Education Committee 14–16 November 1931 [cross in red] An order of service to be used at Gregynog on Sunday 15 November 1931 [cross in red] *Colophon*: Gregynog Press 1931

[2] 11 [3] pp. 285 × 188 mm. Baskerville type on blue cover paper and Abbey Mills machine-made paper. Text printed in black and red. Colophon preceded by Device No. 17.

E87 Order of Service

League of Nations Union An inter-collegiate conference on disarmament 11–14 December 1931 [cross] An order of service to be used at Gregynog on Sunday 13 December 1931 [cross] *Colophon*: Gregynog Press 1931

[2] 11 [3] pp. 285 × 188 mm. Bembo type on light brown cover paper and Abbey Mills machine-made paper. Colophon preceded by Device No. 17.

E88 Christmas Card

With greetings and best wishes for [in red] Christmas and the New Year 1931 – 1932 [in black] from Mrs. E. Davies The Misses Davies Broneirion Gregynog

Single leaf. 88 × 168 mm. Poliphilus type on grey mottled card with deckle edge.

E89 Christmas Card

[In red] 1931 – 1932 [in black] Greetings and best wishes for Christmas and the New Year [Device No. 18 in red] from the Misses Davies Gregynog

[6] pp. 267 × 219 mm. Bembo type on cream cover paper and Japanese vellum. The covers enclose a loose leaf, measuring 255 × 210 mm, of 22 lines from John Drinkwater's *The Fires of God*, printed in red and black on Japanese vellum. The two verses begin with large wood engraved initials T and O, in red.

E90 Christmas Card

[In red] 1931 – 1932 [in black] Greetings and best wishes for Christmas and the New Year [Device No. 18 in red] from The Misses Davies Gregynog

[6] pp. 265 × 215 mm. Bembo type on blue cover paper and hand-made paper. The covers enclose a loose leaf, measuring 255 × 210 mm, of 14 lines from Spenser's *Hymn to eternall love*, printed in red and black on hand-made paper. The lines open with a large wood engraved initial B, in red.

1932

E91 Order of Service

Distressed Areas of South Wales & Monmouthshire Conference on Social Reconstruction 15–18 January 1932 [cross] An order of service to be used at Gregynog on Sunday 17 January 1932 [cross] *Colophon*: Gregynog Press 1932
[4] 4 [2] pp. 285 × 188 mm. Garamond type on brown cover paper and Abbey Mills machine-made paper. On final text recto, 'Printed for private use at Gregynog'. Colophon preceded by Device No. 17.

E92 Prospectus

[Engraved title] The fables of Esope translated out of Frensshe in to Englysshe by : William Caxton with engravings on wood by Agnes Miller Parker and initial letters by William MacCance . . . The Gregynog Press: Newtown, Montgomeryshire: – MCMXXXII.
[4] pp. 305 × 211 mm. Bembo type on hand-made paper. Contains a description of the edition followed by Device No. 18, two specimen pages and lists six books issued 1929–1931.

E93 Concert Programme

A concert of vocal and instrumental music [ornament] Gregynog 11th April 1932 *Colophon*: The Gregynog Press 1932 [Device No. 16]
[8] pp. 285 × 190 mm. Poliphilus type on Abbey Mills machine-made paper. Printed on one side of the leaf only.

E94 Dedication

To Adrian Boult with gratitude from the members of the Gregynog Choir 11th April, MCMXXXII "One man with a dream, at pleasure, / Shall go forth and conquer a crown; / And three with a song's new measure / Can trample a kingdom down."
[4] pp. 255 × 178 mm. Poliphilus type on Japanese vellum. On inside recto, 'Benedicite for soprano solo, chorus and orchestra The words taken from "The song of the three Holy children" and a poem by J. Austin (1613–1669) and set to music by R. Vaughan Williams'. This was the last item to be performed at the concert on 11 April.

E95 The League of Nations Union. Welsh Council. World Wireless Message

The eleventh annual wireless message of the youth of Wales Unfed – neges – ar ddeg ieuenctid Cymru trwy'r radio Printed at the Gregynog Press
Single leaf. 315 × 415 mm. Bembo type on hand-made paper. Text in Welsh and English, printed in two columns, divided by a wood engraved dragon, printed in red. Each column is dated at the foot 'May 18, 1932'.

E96 Certificate

[Within a wood engraved chevron border] Eisteddfod Genedlaethol Urdd Gobaith Cymru [in red] Machynlleth [wood engraving] 1932 [in black] Hyn sydd i arwyddo i o ennill clod ac anrhydedd yr eisteddfod yn am Gwasg Gregynog
Single leaf. 354 × 235 mm. Caslon type on hand-made paper. The border measures 170 × 185 mm. The wood engraving is a black outline shield containing a red dragon.

E97 Prospectus
[In red] Caniadau [double rule] gan W. J. Gruffydd . . . [in red] Gwasg Gregynog
Single leaf. 254×183 mm. Perpetua type on hand-made paper. The area between title and imprint is occupied with a description of the edition in Welsh and English.

E98 Invitation
A concert will be given at Gregynog 16th July, 1932 at 8 p.m. The audience is requested to be seated at 7.50 p.m. Please bring this card with you
Single leaf. 89×140 mm. Garamond type on card.

E99 Concert Programme
A concert at Gregynog 16th. July, 1932 [ornament] *Colophon*: The Gregynog Press [Device No. 16]
[8] pp. 255×170 mm. Poliphilus type on Abbey Mills machine-made paper. Printed on one side of the leaf only.

E100 Order of Service
National Council of Music thirteenth annual conference 15–18 July 1932 [cross in red] An order of service to be used at Gregynog on Sunday 17 July 1932 [cross in red] *Colophon*: Gregynog Press 1932
[2] 11 [3] pp. 278×180 mm. Perpetua type on light brown cover paper and Abbey Mills machine-made paper. Text printed in black and red. On final text recto, 'Printed for private use at Gregynog'. Colophon preceded by Device No. 17.

E101 Exhibition Piece
The sorrows of young Werther. Translated by Dr. William Rose and published originally by the Scholartis Press, London. *Colophon*: Pages reprinted and decorated by William MacCance at the Gregynog Press, Newtown, Montgomeryshire for the Goethe exhibition at Leipzig. Hand-set by Idris Jones. Pressman: Herbert John Hodgson. MCMXXXII
[4] pp. 265×167 mm. Perpetua type on Japanese vellum. Consists of pages 84 and 85, printed in black and reddish-brown. Page 84 opens with a wood engraved decorative initial O printed in the two colours.

E102 Order of Service
In memory of Annie Hughes died 25th August 1932 An order of service to be used at Gregynog on Sunday 28 August 1932 *Colophon*: Gregynog Press 1932
[2] 7 [3] pp. 278×180 mm. Perpetua type on brown cover paper and Abbey Mills machine-made paper. On final text verso, 'Printed for private use at Gregynog'. Colophon preceded by Device No. 17. Annie Hughes was a housemaid at Gregynog Hall.

E103 Prospectus
The singing caravan[:] a Sufi tale by Robert Vansittart with initial letters and frontispiece in two colours by William MacCance . . .
Single leaf. 234×150 mm. Perpetua type on hand-made paper. Includes a description of the edition and a wood engraving printed in reddish-brown ink.

E104 Prospectus

Erewhon by Samuel Butler The wood engravings by Blair Hughes-Stanton . . . MCMXXXII.

[4] pp. 232 × 154 mm. Baskerville type on Japanese vellum. Consists of a description of the edition, two specimen pages from *Erewhon* and lists *The Revelation of Saint John the Divine*, *Comus*, *The Fables of Esope*, *Caniadau gan W. J. Gruffydd* and *The Singing Caravan*.

E105 Order of Service

League of Nations Union Welsh National Council Advisory Education Committee 18–21 November 1932 [cross in red] An order of service to be used at Gregynog on Sunday 20 November 1932 [cross in red] *Colophon*: Gregynog Press 1932

[2] 15 [3] pp. 278 × 180 mm. Perpetua type on light brown cover paper and Abbey Mills machine-made paper. Text printed in red and black. On final text verso, 'Printed for private use at Gregynog'. Colophon preceded by Device No. 17.

E106 Order of Service

South Wales Girls' Camp 3–5 December 1932 An order of service to be used at Gregynog on Sunday 4 December 1932 *Colophon*: Gregynog Press 1932

[2] 10 pp. 278 × 180 mm. Perpetua type on brown cover paper and Abbey Mills machine-made paper. On final text verso, 'Printed for private use at Gregynog'. Colophon preceded by Device No. 17.

E107 Christmas Card

[In red] To bring greetings and best wishes for Nineteen Thirty Three [in black] from the Misses Davies [in red] MCMXXXII *Colophon*: The Gregynog Press 1932

[6] pp. 271 × 195 mm. Poliphilus type on grey silurian cover paper, Blado italic on hand-made paper. The covers enclose a loose leaf, measuring 255 × 190 mm, of 11 lines by Lascelles Abercrombie beginning, 'This, then, is yours; . . .' printed in red and black which opens with a large wood engraved initial T. Colophon preceded by Device No. 16.

E108 Christmas Card

[Wood engraving incorporating BC]

[4] pp. 125 × 173 mm. Perpetua type on heavy Japanese vellum. The wood engraving incorporating BC (Bovington Camp) depicts a tent, sand dunes and the sea. On inside recto, 'To bring you all that is best for Christmas and the New Year 1932 from Margaret S. Davies Gregynog, Newtown'.

1933

E109 Order of Service

Distressed Areas of South Wales Conference on Social Reconstruction 27 – 30 January 1933 [cross in red] An order of service to be used at Gregynog on Sunday 29 January 1933 [cross in red] *Colophon*: Gregynog Press 1933

[2] 18 pp. 278 × 180 mm. Perpetua type on brown cover paper and Abbey Mills machine-made paper. Text printed in red and black. On final text verso, 'Printed for private use at Gregynog'. Colophon preceded by Device No. 17.

E110 Invitation

[Wood engraving] The honour of your company is requested at a private view on Wednesday April 5 1933 of an exhibition of the wood engravings of Blair Hughes-Stanton including those engraved for The Revelation of Saint John published by the Gregynog Press at the Zwemmer Gallery 26 Litchfield Street, Charing Cross Rd. W.C.2 April 5 to April 29 10 a.m. to 6 p.m.

Single leaf. 155 × 115 mm. Perpetua type on heavy Japanese vellum. On the reverse side is a sketch map showing the location of the Zwemmer gallery.

E111 Prospectus

The Revelation of Saint John [in red] with engravings by Blair Hughes-Stanton . . .

Single leaf. 140 × 173 mm. Bembo type with Perpetua titling on Japanese vellum. Consists of a description of the edition, printed in red and black.

E112 Prospectus

[Wood engraved title] The Revelation of [in red] Saint John the Divine [in black, Device No. 22] The Gregynog Press [in red] MCMXXXIII

[4] pp. 345 × 205 mm. Bembo type with Perpetua titling on heavy Japanese vellum. Contains two specimen pages and a description of the edition, printed in red and black.

E113 League of Nations Union. Welsh Council. World Wireless Message

The twelfth annual wireless message of the youth of Wales Deuddegfed neges flynyddol ieuenctid Cymru trwy'r radio Printed at the Gregynog Press

Single leaf. 315 × 415 mm. Bembo type on hand-made paper. Text in Welsh and English, printed in two columns, divided by a wood engraved dragon printed in red. Each column is dated at the foot, 'May 18, 1933'.

E114 Invitation

You are invited to a Festival of Music and Poetry to be held at Gregynog June 23rd to 26th 1933 . . . . Please reply *as soon as possible* on the enclosed card marking which concerts you may wish to attend

[4] pp. 204 × 124 mm. Perpetua type on Japanese vellum. Contains a list of artists and a list of concerts to be held.

E115 Reply Card

I am happy to accept your invitation for the concert on: Friday . . . . 23 June, 1933 Saturday . . . . 24 June, 1933 Monday . . . . 26 June, 1933 Name and address: I regret that I am unable to accept your invitation. Name and address:

Single leaf. 95 × 140 mm. Perpetua type on card.

E116 Festival Programme

Festival of Music and Poetry Gregynog June 23rd to 26th 1933 [Device No. 23] In aid of the Coleg Harlech Training Centre for Unemployed *Colophon*: Gregynog Press 1933

[6] 20 [2] pp. 282 × 187 mm. Perpetua type on cream cover paper and Abbey Mills machine-made paper. Title repeated on cover, set in a pattern of rules and square dots. Complete programme including libretti. Colophon preceded by Device No. 23.

E117 Festival Programme
Festival of Music and Poetry Gregynog Friday June 23rd 1933 [Device No. 23] In aid of the Coleg Harlech Training Centre for Unemployed *Colophon*: The Gregynog Press 1933
[4] pp. 282 × 187 mm. Perpetua type on Abbey Mills machine-made paper. Colophon preceded by Device No. 24.

E118 Festival Programme
Festival of Music and Poetry Gregynog Saturday June 24th 1933 [Device No. 23] In aid of the Coleg Harlech Training Centre for Unemployed *Colophon*: The Gregynog Press 1933
[4] pp. 282 × 187 mm. Perpetua type on Abbey Mills machine-made paper. Colophon preceded by Device No. 24.

E119 Festival Programme
Festival of Music and Poetry Gregynog Sunday June 25th 1933 [Device No. 23] In aid of the Coleg Harlech Training Centre for Unemployed *Colophon*: The Gregynog Press 1933
[4] pp. 282 × 187 mm. Perpetua type on Abbey Mills machine-made paper. Colophon preceded by Device No. 24.

E120 Festival Programme
Festival of Music and Poetry Gregynog Monday June 26th 1933 [Device No. 23] In aid of the Coleg Harlech Training Centre for Unemployed *Colophon*: The Gregynog Press 1933
[4] pp. 282 × 187 mm. Perpetua type on Abbey Mills machine-made paper. Colophon preceded by Device No. 24.

E121 Order of Service
Festival of Music and Poetry 23–26 June 1933 An order of service to be used at Gregynog on Sunday 25 June 1933 *Colophon*: Gregynog Press 1933
[6] 18 [4] pp. 283 × 188 mm. Perpetua type on stone cover papers and Abbey Mills machine-made paper. Colophon preceded by Device No. 17.

E122 Acknowledgment
Gregynog, Newtown, Montgomeryshire July, 1933 The sum of £112 has been sent to the Treasurer, Coleg Harlech, for the training of unemployed men, as the result of the appeal made at the Gregynog Festival of Music and Poetry, 23rd – 26th June, 1933. Gwendoline and Margaret Davies would like to thank all whose generous help and co-operation made the spirit of this festival an unforgettable memory.
Single leaf. 203 × 127 mm. Baskerville italic type on Japanese vellum.

E123 Order of Service
National Council of Music fourteenth annual conference 14 – 27 July 1933 An order of service to be used at Gregynog on Sunday 16 July 1933 *Colophon*: Gregynog Press 1933
[4] 9 [3] pp. 285 × 185 mm. Perpetua type on brown cover paper and Abbey Mills machine-made paper. On penultimate text verso, 'Printed for private use at Gregynog'. Colophon preceded by Device No. 17.

E124 Prospectus

[In brown] Clych atgof [double rule in brown] [in black] gan: [in brown] Owen Edwards . . . Gwasg Gregynog ger y Drenewydd Sir Drefaldwyn MCMXXXIII
Single leaf. 213 × 141 mm. Perpetua type on hand-made paper. Consists of a description of the edition in Welsh and English.

E125 Certificate

[Within a wood engraved chevron border] Eisteddfod Genedlaethol Urdd Gobaith Cymru [in red] Caerffili [wood engraving] 1933 [in black] Hyn sydd i arwyddo i o ennill clod ac anrhydedd yr eisteddfod yn am Gwasg Gregynog
Single leaf. 285 × 355 mm. Caslon type on hand-made paper. The border measures 170 × 185 mm. The wood engraving is a black outline shield containing a red dragon. 700 copies printed.

E126 Invitation

Gregynog, Newtown, Montgomeryshire You are invited to view an exhibition of work done in the distressed areas of South Wales and at other centres, to be opened by Mrs. Wintringham, J. P., at Gregynog on Wednesday, 30th August, 1933 at 3.0 p.m. Representatives from the various centres will be present and will give short accounts of the recent developments in craftwork & in social reconstruction in these areas. Articles may be purchased and orders given. R.S.V.P. The Secretary, Gregynog
Single leaf. 203 × 127 mm. Baskerville type on Japanese vellum.

E127 Exhibition Opening Programme

Gregynog, Newtown, Montgomeryshire. Wednesday, 30th August, 1933 Opening of exhibition by Mrs. Wintringham, J. P. The Rhondda Valley Industries, and Maes-yr-Haf Settlement, Trealaw. Emma Noble & William Noble. Brynmawr and Clydach Valley Industries. Peter Scott, Paul Matt & Noel Hyde. Merthyr Quilters. Margaret Gardner. South Wales Girls' Clubs. Margaret George. Unemployed Men's Centre, Newtown. Ernest Evans & Clement Evans. An exhibit of handwork will be on view from the Children's Home, Ty Gwyn, Llwyngwril.
Single leaf. 225 × 130 mm. Baskerville type on Japanese vellum.

E128 Pamphlet

[In blue] The Blue Pilgrims *Colophon*: Printed at the Gregynog Press 1933
[12] pp. 250 × 177 mm. Bembo type on grey cover paper and hand-made paper. Text opens with the word 'Who' in blue of which the initial letter is wood engraved. An account of the social service movement initiated by Beatrice Hankey. Title repeated in blue on cover. Colophon preceded by Device No. 7 in blue. 1,000 copies printed.

E129 Prospectus

Four poems by John Milton L'Allegro Il Penseroso Arcades Lycidas [wood engraving] With wood engravings by Blair Hughes-Stanton . . .
Single leaf. 233 × 138 mm. Perpetua type on Japanese vellum. Consists of a description of the edition.

E130 Prospectus

XXI Welsh gypsy folk-tales collected by John Sampson [wood engraving] With engravings on wood by Agnes Miller Parker . . .
[4] pp. 275 × 195 mm. Bembo type on hand-made paper. Includes a description of the edition, a specimen page and a list of books recently issued.

E131 Order of Service

League of Nations Union Welsh National Council Advisory Education Committee 24 – 27 November 1933 An order of service to be used at Gregynog on Sunday 26 November 1933 *Colophon*: Printed for private use at Gregynog
[4] 15 [5] pp. 285 × 188 mm. Baskerville type on stone cover paper and Abbey Mills machine-made paper. On lower cover, Device No. 12 and MCMXXXIII.

E132 Order of Service

South Wales Girls' Camp 8 [altered in ink to 9] – 11 December 1933 An order of service to be used at Gregynog on Sunday 10 December 1933 *Colophon*: Printed for private use at Gregynog
[2] 8 [2] pp. 280 × 180 mm. Perpetua type on brown cover paper and Abbey Mills machine-made paper. On lower cover, Device No. 12 and MCMXXXIII.

E133 Prospectus

The Gregynog Press announces the issue of thirty new poems by W. H. Davies The lovers' song-book . . .
Single leaf. 170 × 140 mm. Baskerville type on Japanese vellum. Consists of a description of the edition.

E134 Prospectus

The lovers' song-book [double rule in green] New poems by W. H. Davies . . .
Single leaf. 212 × 140 mm. Baskerville type on Japanese vellum. Consists of a description of the edition.

E135 Prospectus

[Facsimile of wood engraved title-page in blue and black] The lamentations of Jeremiah . . . The Gregynog Press MCMXXXIII.
[4] pp. 385 × 258 mm. Baskerville type on heavy Japanese vellum. Includes a description of the edition and two specimen pages, printed in blue and black.

E136 Christmas Card

To bring greetings and best wishes for Christmas and the New Year 1933 – 1934 [asterisk] from the Misses Davies Gregynog
[6] pp. 265 × 215 mm. Poliphilus type on grey silurian cover paper, Poliphilus and Blado italic on hand-made paper. The covers enclose a loose leaf, measuring 260 × 210 mm of lines beginning, 'The future lies large and splendid before us . . .', printed in red and black which opens with a large wood engraved initial T.

1934

E137 Order of Service
Distressed Areas of South Wales Conference on Social Reconstruction 2 – 5 February 1934 [cross in red] An order of service to be used at Gregynog on Sunday 4 February 1934 [cross in red] *Colophon*: Printed for private use at Gregynog
[2] 16 [2] pp. 285 × 187 mm. Perpetua type on stone cover papers and Abbey Mills machine-made paper. Text printed in red and black. On lower cover, Device No. 12 and MCMXXXIV.

E138 League of Nations Union. Welsh Council. World Wireless Message
Neges flynyddol ieuenctid [wood engraved numerals in red occupying two lines of type] XIII Annual wireless message Cymru trwy'r radio i'r byd of the youth of Wales Printed at the Gregynog Press
Single leaf. 270 × 330 mm. Perpetua type on hand-made paper. Text in Welsh and English, printed in two columns, divided by a wood engraved dragon printed in red. Each column is dated at the foot, 'May 18, 1934'.

E139 Certificate
[Within a wood engraved chevron border] Eisteddfod Genedlaethol Urdd Gobaith Cymru [in red] Hen Golwyn [wood engraving] [in red] 1934 [in black] Hyn sydd i arwyddo i o ennill clod ac anrhydedd yr eisteddfod yn am
Single leaf. 285 × 355 mm. Caslon type on hand-made paper. The border measures 170 × 185 mm. The wood engraving is a black outline shield containing a red dragon.

E140 Festival Programme
The second Festival of Music and Poetry at Gregynog 15 – 18 June 1934 [Device No. 23] Contributions in aid of the unemployed will be accepted at the door. *Colophon*: The Gregynog Press 1934
[6] 36 [2] pp. 284 × 190 mm. Perpetua type on grey silurian paper covers and Abbey Mills machine-made paper. Title repeated on cover, set in a pattern of rules and square dots. Complete festival programme including libretti. Colophon preceded by Device No. 24.

E141 Festival Programme
The second Festival of Music and Poetry at Gregynog Friday, 15th June 1934 [Device No. 23 in blue] Contributions in aid of the unemployed will be accepted at the door. *Colophon*: Printed at the Gregynog Press Montgomeryshire 1934
[4] pp. 284 × 190 mm. Perpetua type on Abbey Mills machine-made paper.

E142 Festival Programme
The second Festival of Music and Poetry at Gregynog Saturday, 16th June 1934 [Device No. 23 in blue] Contributions in aid of the unemployed will be accepted at the door. *Colophon*: Printed at the Gregynog Press Montgomeryshire 1934
[4] pp. 284 × 190 mm. Perpetua type on Abbey Mills machine-made paper.

E143 Festival Programme

The second Festival of Music and Poetry at Gregynog Sunday 17th June 1934 [Device No. 23 in blue] Contributions in aid of the unemployed will be accepted at the door. *Colophon*: Printed at the Gregynog Press Montgomeryshire 1934

[6] pp. 284 × 190 mm. Perpetua type on Abbey Mills machine-made paper. Consists of a single fold enclosing a single leaf, measuring 285 × 181 mm, on which is printed the libretto for Vaughan Williams's *Mass in G. Minor.*

E144 Festival Programme

The second Festival of Music and Poetry at Gregynog Monday 18th June 1934 [Device No. 23 in blue] Contributions in aid of the unemployed will be accepted at the door. *Colophon*: Printed at the Gregynog Press Montgomeryshire 1934

[4] pp. 284 × 190 mm. Perpetua type on Abbey Mills machine-made paper.

E145 Poem

An ode written in honour of Sir Walford Davies C.V.O. [in blue] Master of the King's Musick [in black] by Lascelles Abercrombie and read by the author at the second Gregynog Festival of Music and Poetry [in blue] MCMXXXIV *Colophon*: Printed at the Gregynog Press Montgomeryshire 1934

[8] pp. 280 × 194 mm. Poliphilus type on grey silurian covers and hand-made paper. On the front cover, Device No. 12 and 'Gregynog 1934'. Text printed in blue and black.

E146 Poem

[In blue] To Sir Walford Davies at Gregynog June, 1934

Single leaf. 420 × 240 mm. Poliphilus type on Japanese vellum. Consists of the ode written by Lascelles Abercrombie, printed in blue and black, which opens with a large wood engraved F.

E147 Order of Service

The second Festival of Music and Poetry 15 – 18 June 1934 [cross in red] An order of service to be used at Gregynog on Sunday 17 June, 1934 [cross in red] *Colophon*: Printed for private use at Gregynog

[4] 16 [4] pp. 280 × 187 mm. Perpetua type on stone cover paper and Abbey Mills machine-made paper. Text printed in red and black. On the lower cover, Levavi [Device No. 12] oculus Gregynog.

E148 Order of Service

National Council of Music fifteenth annual conference 13 – 16 July 1934 [cross in red] An order of service to be used at Gregynog on Sunday 15 July 1934 [cross in red] *Colophon*: Printed for private use at Gregynog

[6] 12 [2] pp. 280 × 187 mm. Perpetua type on grey cover paper and Abbey Mills machine-made paper. Text printed in red and black. On lower cover, Levavi [Device No. 12] oculos Gregynog. Loosely inserted is a folded sheet, measuring 280 × 187 mm, containing, on the inner verso, the words of *O spiritual pilgrim* by Gustav Holst and on the inner recto, the words of *Ah! gentle may I lay me down* by H. Walford Davies. On the front, 'National Council of Music fifteenth annual conference Sunday 15th July 1934'.

E149 Prospectus
The Gregynog Press [Device No. 10, white on blue] July 1934
[4] pp. 240 × 170 mm. Baskerville type on hand-made paper. The inner pages consist of a description of the edition of *Don Quixote* and two specimen pages. On final verso, an announcement of forthcoming books. Text printed in blue and black.

E150 Order of Service
League of Nations Union Welsh National Council Advisory Education Committee 16 – 19 November 1934 [cross in red] An order of service to be used at Gregynog on Sunday 18th November 1934 [cross in red] *Colophon*: Gregynog 1934
[4] 15 [5] pp. 285 × 187 mm. Perpetua type on grey cover paper and Abbey Mills machine-made paper. On penultimate text verso, 'Printed for private use at Gregynog'. Colophon preceded by Device No. 23.

E151 Prospectus
Two specimen pages from [in red] Anne Boleyn and other poems by Loyd Haberly [in black] decorated with many coloured capitals drawn by Graily Hewitt . . .
[4] pp. 210 × 170 mm. Bembo type on hand-made paper. Consists of a description of the edition, two specimen pages and an announcement of two forthcoming books, printed in red and black with red and green wood engraved initials.

E152 Order of Service
South Wales Girls' Camp 1 – 3 December 1934 [cross in red] An order of service to be used at Gregynog on Sunday 2nd December 1934 [cross in red] *Colophon*: Gregynog 1934
[2] 8 [2] pp. 285 × 187 mm. Perpetua type on grey cover paper and Abbey Mills machine-made paper. Text printed in red and black. On final text recto, 'Printed for private use at Gregynog'. Colophon preceded by Device No. 23.

E153 Christmas Card
For Nineteen [wood engraved initial G in red occupying four lines of type] reetings and Thirty four best wishes for and Nineteen Christmas & Thirty five the New Year from Gwendoline & Margaret Davies Gregynog
[6] pp. 307 × 228 mm. Poliphilus type on stone cover paper. Covers enclose a loose sheet consisting of a woodcut depicting a windmill signed 'M S Davies' which measures 120 × 116 mm mounted on a sheeting measuring 297 × 223 mm.

E154 Christmas Card
For Nineteen [wood engraved initial G in green occupying four lines of type] reetings and Thirty four best wishes for and Nineteen Christmas & Thirty five the New Year from Margaret S. Davies Gregynog
[4] pp. 215 × 165 mm. Poliphilus type on hand-made paper. On inside recto, a wood engraved holly branch followed by two verses of *The holly and the ivy*, printed in green and black.

1935

E155 Order of Service
Distressed Areas of South Wales Conference on Social Reconstruction 1–4 February

1935 [cross in red] An order of service to be used at Gregynog on Sunday 3 February 1935 [cross in red] *Colophon*: Gregynog 1935
[4] 17 [3] pp. 288 × 188 mm. Perpetua type on grey cover paper and Abbey Mills machine-made paper. Text printed in red and black. On final text verso, 'Printed for private use at Gregynog February 1935'. Colophon preceded by Device No. 23.

E156 Anthem
[In red] In thy strength, O Lord an anthem [in black] words from Canon Dalton's translation of the Psalms with music by Walford Davies [in red] Sunday, 5 May, 1935
[4] pp. 290 × 205 mm. Garamond type on hand-made paper. On inside recto, a crown in red and 'Humbly dedicated to the King by His Majesty's gracious permission'. Anthem performed in Saint Paul's Cathedral. Two copies printed on vellum for the use of the King and Queen.

E157 The League of Nations Union. Welsh Council. World Wireless Message
Dydd Ewyllys Da [wood engraved numerals] XIV Goodwill Day Annual world wireless message of the youth of Wales Neges radio flynyddol plant Cymru i'r byd Printed at the Gregynog Press, May MCMXXXV
Single leaf. 292 × 380 mm. Bembo type on hand-made paper. Text in Welsh and English, printed in two columns divided by a wood engraved dragon printed in red.

E158 Prospectus
The star of Seville: a drama in three acts and in verse attributed to Lope de Vega translated out of Spanish by Henry Thomas [wood engraving in yellow, red, blue and black] Gregynog MCMXXXV
[4] pp. 247 × 170 mm. Bembo type on hand-made paper. Printed in black, red and green. The wood engraving is the coat of arms of Seville. Item contains a specimen page and a description of the edition together with a description of *Don Quixote*, already published.

E159 Certificate
[Within a wood engraved chevron border] Eisteddfod Genedlaethol Urdd Gobaith Cymru [in red] Caerfyrddyn [wood engraving] [in red] 1935 [in black] Hyn sydd i arwyddo i o ennill clod ac anrhydedd yr eisteddfod yn am Gwasg Gregynog
Single leaf. 285 × 355 mm. Caslon type on hand-made paper. The border measures 170 × 185 mm. The wood engraving is a black outline shield containing a red dragon.

E160 Festival Programme
The third Festival of Music & Poetry to be held at Gregynog June 28th–July 1st 1935 [Device No. 23] Contributions in aid of the Orthopaedic Hospital and Derwen Cripples Training College Oswestry will be accepted *Colophon*: Set within a pattern of rules, Printed at the Gregynog Press June, 1935
[8] 33 [3] pp. 284 × 189 mm. Perpetua type on grey silurian paper and Abbey Mills machine-made paper. Title repeated on cover, set in a pattern of rules. On penultimate text verso, 'Printed for private use at Gregynog, June, 1935'.

E161 Appeal
The third Gregynog Festival of Music and Poetry June 28th – July 1st 1935 The Robert Jones & Agnes Hunt Orthopaedic Hospital and Derwen Cripples Training College Oswestry Contributions will be gratefully accepted during the interval each evening
[4] pp. 210×132 mm. Baskerville type on hand-made paper. Printed on one side of the sheet only. Contains a description of the two institutions and an appeal for funds.

E162 Order of Service
Third Festival of Music and Poetry June 28th – July 1st, 1935 [cross in red] An order of service to be used at Gregynog on Sunday, 30th June 1935 [cross in red] *Colophon*: Gregynog 1935
[12] pp. 285×190 mm. Perpetua type on grey cover paper and Abbey Mills machine-made paper. Text printed in red and black. On penultimate text verso, 'Printed for private use at Gregynog'. The colophon is preceded by Device No. 23.

E163 Prospectus
Specimen pages from Eros and Psyche [wood engraving] A poem in XII measures by Robert Bridges: with wood-cuts from designs by Edward Burne-Jones Gregynog MCMXXXV
[4] pp. 282×217 mm. Gregynog type on hand-made paper. Consists of two specimen pages and a description of the edition.

E164 Order of Service
League of Nations Union Welsh National Council Advisory Education Committee 22–25 November 1935 [cross in red] An order of service to be used at Gregynog on Sunday November 24 1935 [cross in red] *Colophon*: Gregynog 1935
[4] 13 [3] pp. 280×190 mm. Perpetua type on stone cover paper and Abbey Mills machine-made paper. Text printed in red and black. On final text verso, 'Printed for private use at Gregynog, November 1935'. Colophon preceded by Device No. 23.

E165 Order of Service
South Wales Girls' Camp 7–9 December 1935 An order of service to be used at Gregynog on Sunday December 8 1935 *Colophon*: Gregynog 1935
[2] 7 [3] pp. 280×190 mm. Perpetua type on stone cover paper and Abbey Mills machine-made paper. On penultimate text verso, 'Printed for private use at Gregynog'. Colophon preceded by Device No. 23.

E166 Poem
New York City by Helen Waddell with a wood-cut impression by Stefan Mrozewski *Colophon*: Printed under the direction of Stefan Mrozewski at the Gregynog Press December 5 1935
[16] pp. 280×190 mm. Poliphilus and Blado italic types on grey cover paper and hand-made paper. Title repeated on cover. Woodcut frontispiece signed in pencil by the artist. On penultimate text recto, '350 copies of this poem and engraving have been printed. Of 100 signed impressions this is number ...'. Numbered and signed in pencil by the poet.

E167 Christmas Card

New York City by Helen Waddell with a wood-cut impression by Stefan Mrozewski *Colophon*: Printed under the direction of Stefan Mrozewski at the Gregynog Press December 5 1935

[16] pp. 276×178 mm. Poliphilus and Blado italic types on cream cover paper and hand-made paper. Cover title 'To bring greetings and best wishes for Nineteen thirty six from Gregynog Christmas 1935'. On final text recto, '350 copies of this poem and engraving have been printed'.

Variant copies have blue-grey covers.

E168 Christmas Card

An old Welsh carol With greetings and best wishes for Christmas and the New Year from Gwendoline and Margaret Davies Gregynog 1935–1936 *Colophon*: Privately printed at Gregynog under the direction of Loyd Haberly who designed these types for the Press.

[8] pp. 255×297 mm. Gregynog type and special music type, designed by Loyd Haberly, on hand-made paper. Plain grey covers. The text consists of the carol, *Cwyd, a gwel yr adeilad* and a free translation of the words, printed in red, green and black. There are three small wood engravings and a wood engraved initial.

1936

E169 Order of Service

Distressed Areas of South Wales Conference on Social Reconstruction 7–10 February 1936 [cross in red] An order of service to be used at Gregynog on Sunday February 9 1936 [cross in red] *Colophon*: Gregynog 1936

[2] 12 [2] pp. 282×190 mm. Perpetua type on grey cover paper and Abbey Mills machine-made paper. Text printed in red and black. On final text verso, 'Printed for private use at Gregynog, February 1936'. Colophon preceded by Device No. 12.

E170 League of Nations Union. Welsh Council. Annual Wireless Message

Pymthegfed neges [wood engraving in red] Fifteenth annual flynnyddol plant cymru world wireless message trwy'r radio i'r byd of the youth of Wales Dydd Ewyllys Da Goodwill Day Printed at the Gregynog Press, May MCMXXXVI

Single leaf. 290×375 mm. Garamond type on hand-made paper. The wood engraving is a dragon which occupies four lines of type and divides Welsh and English titles. Text in Welsh and English, printed in two columns.

E171 Certificate

[Within a wood engraved chevron border] Eisteddfod Genedlaethol Urdd Gobaith Cymru 4–6 Mehefin 1.9.3.6 Blaenau [wood engraving] Ffestiniog Hyn sydd i arwyddo i o ennill clod ac anrhydedd yr eisteddfod yn am Gwasg Gregynog

Single leaf. 320×274 mm. Caslon type on hand-made paper. The border measures 170×185 mm. The wood engraving is a black outline shield containing a red dragon.

E172 Invitation

You are invited to the fourth Festival of Music and Poetry to be held at Gregynog July, 3–5th 1936 Contributions in aid of the Orthopaedic Hospital and Derwen

Cripples Training College, Oswestry will be accepted Please send an early reply [rule]
[4] pp. 180 × 128 mm. Perpetua type on hand-made paper. Contains a list of events, a list of chief artists and the appeal notice. Signed in ink by Gwendoline E. Davies and Margaret S. Davies.

E173 Invitation
The fourth Festival of Music and Poetry to be held at Gregynog 1936 Contributions in aid of the Orthopaedic Hospital and Derwen Cripples Training College, Oswestry will be accepted
[4] pp. 182 × 120 mm. Perpetua type on hand-made paper. On inside verso, 'You are invited to the Festival concert to be given in the Music Room – Gregynog on Saturday, July 4, at 8 p.m.'. This is followed by a list of artists and a note concerning the charity.

E174 Festival Programme
The fourth Festival of Music and Poetry at Gregynog 3–5, July 1936 [Device No. 23] Contributions in aid of the Orthopaedic Hospital and Derwen Cripples Training College, Oswestry will be accepted *Colophon*: The Gregynog Press July, 1936
[6] 33 [1] pp. 285 × 190 mm. Perpetua type on blue cover paper and Abbey Mills machine-made paper. Title repeated on cover, set in a pattern of rules and square dots. On final text verso, 'Printed for private use at Gregynog, July 1936'. Colophon preceded by Device No. 24.

E175 Order of Service
The fourth Festival of Music and Poetry 3–5 July 1936 [cross in red] An order of service to be used at Gregynog on Sunday, 5 July 1936 [cross in red] *Colophon*: Gregynog July, 1936
[4] 16 [4] pp. 280 × 187 mm. Perpetua type on double grey cover paper and hand-made paper. Text printed in red and black. On penultimate text verso, 'Printed for private use at Gregynog, July 1936'. Colophon preceded by Device No. 12 dividing the motto 'Levavi oculos'.

E176 Review
The Gregynog Festival 1936 by A. K. Holland from The Liverpool Daily Post of August 8, 1936
[4] pp. 190 × 120 mm. Perpetua type on Abbey Mills machine-made paper. A reprint of the original review.

E177 Prospectus
Specimen pages from Cyrupaedia: the institution and life of Cyrus, the first of that name, King of Persians. Eight bookes treating of noble education, of princely exercises, military discipline, warlike stratagems, preparations & expeditions: written in Greeke by the sage Xenophon. [in red] Translated out of Greeke into English, and conferred with the Latine and French translations, by Philemon Holland of the city of Coventry doctor in physick. [wood engraving in black] [in red] Dedicated to his most excellent Majesty. [in black] First printed by J. L. for Robert Allot, to be sold at the Sign of the Beare in Paul's Churchyard, 1632, and now fully

reprinted from the 1632 edition at the Gregynog Press, Newtown, Mont. MDCCCCXXXVI
[4] pp. 305 × 215 mm. Poliphilus type on hand-made paper. The wood engraving is of the royal crown. Contains two specimen pages, a note that, 'his book has been reprinted . . . under the direction of Mr. Loyd Haberly, who has edited the text in collaboration with Mr. F. W. Bateson', and a description of the edition. Printed in red and black.

E178 Order of Service
National Council of Music seventeenth annual conference [cross in red] An order of service to be used at Gregynog on Sunday October 4 1936 [cross in red] *Colophon*: Gregynog October, 1936
[2] 13 [1] pp. 280 × 188 mm. Perpetua type on double grey cover paper and Abbey Mills machine-made paper. Text printed in red and black. On final text verso, 'Printed for private use at Gregynog, October 1936'. Colophon preceded by Device No. 12.

E179 Prospectus
[In red] The story of a red-deer [in black] by the Honourable J. W. Fortescue With decorations by Dorothy Burroughes [wood engraving in green] [in black] The Gregynog Press MCMXXXVI
[4] pp. 250 × 195 mm. Cloister type on hand-made paper. Includes a specimen page, a note about the book and a description of the edition.

E180 Prospectus
The Gregynog Press and its books
[4] pp. 220 × 157 mm. Perpetua type on hand-made paper. Consists of a note about the Press and a list of the first thirty-four books, seventeen of which are asterisked as being out of print. *The story of a red-deer* is erroneously ascribed to 1935.

E181 Order of Service
In memory of Richard Ingram died 19th November, 1936 An order of service to be used at Gregynog on Saturday November 21 1936 *Colophon*: Gregynog November 1936
[2] 4 [2] pp. 280 × 188 mm. Perpetua type on grey cover paper and Abbey Mills machine-made paper. Colophon preceded by Device No. 23. Richard Ingram was the estate carpenter.

E182 Order of Service
League of Nations Union Welsh National Council Advisory Education Committee 21–23 November 1936 [cross in red] An order of service to be used at Gregynog on Sunday November 22 1936 [cross in red] *Colophon*: Gregynog November, 1936
[4] 14 [2] pp. 280 × 189 mm. Perpetua type on double grey cover paper and Abbey Mills machine-made paper. Text printed in red and black. On final text verso, 'Printed for private use at Gregynog, November 1936'. Colophon preceded by Device No. 23.

E183 Order of Service
South Wales Girls' Camp 5–7 December 1936 An order of service to be used at Gregynog on Sunday 6 December 1936 *Colophon*: Gregynog December 1936

[2] 8 [2] pp. 277 × 187 mm. Perpetua type on grey cover paper and Abbey Mills machine-made paper. On final text verso, 'Printed for private use at Gregynog, December 1936'. Colophon preceded by Device No. 23.

E184 Christmas Card
[In red] With greetings and best wishes for Christmas and the New year [in black] from The Gregynog Press Newtown, Montgomeryshire [in red] 1936 1937
[8] pp. 234 × 155 mm. Bembo and Bembo Condensed Italic type on hand-made paper. Text printed in red and black, on one side of the sheet only. The inner pages consist of G. K. Chesterton's poem *The wise men.*

1937

E185 Order of Service
Distressed Areas of South Wales Conference on Social Reconstruction 5–8 February 1937 An order of service to be used at Gregynog on Sunday February 7 1937 *Colophon*: Gregynog February 1937
[4] 14 [2] pp. 278 × 187 mm. Perpetua type on double grey cover paper and Abbey Mills machine-made paper. On final text verso, 'Printed for private use at Gregynog, February 1937'. Colophon preceded by Device No. 23.

E186 Prospectus
The Gregynog Press announces the publication of [in red] Caelica [in black] by Fulke Greville, Lord Brooke, edited, with an introduction, by Una Ellis-Fermor . . .
[4] pp. 235 × 155 mm. Perpetua type on hand-made paper. Printed in black and red. Contains a note on the author and his poems and a specimen page.

E187 League of Nations Union. Welsh Council. World Wireless Message
Dydd Ewyllys Da, 18 Mai XVI Goodwill Day, 18 May Neges flynyddol plant Cymru [wood engraving in red] Annual world wireless message trwy'r radio i'r byd of the youth of Wales Printed at the Gregynog Press, May MCMXXXVII
Single leaf. 282 × 365 mm. Perpetua type on hand-made paper. The wood engraving is a dragon which occupies two lines of type and divides Welsh and English titles. Text in Welsh and English, printed in two columns.

E188 Certificate
[Within a wood engraved chevron border] Eisteddfod Genedlaethol Urdd Gobaith Cymru [in red] Gwaun-cae-Gurwen [in black] 17. 18. 19. Mai 19 [wood engraving] 37 Hyn sydd i arwyddo i o ennill clod ac anrhydedd yr eisteddfod yn am Gwasg Gregynog
Single leaf. 320 × 274 mm. Caslon type on hand-made paper. The border measures 170 × 185 mm. The wood engraving is a black outline shield containing a red dragon.

E189 Presentation
Montgomery County Music Festival [wood engraving in red] To Sir Adrian C. Boult, M.A., D.Mus. from the committee and choirs of the Montgomery County Festival of Music with grateful thanks for his generous and inspiring help as Conductor and Adviser of the Festival during the years [in red] 1925–1937
Single leaf. 403 × 174 mm. Poliphilus and Blado Italic types on grey card. The wood

engraving, which is circular and measures 50 mm in diameter, depicts the county arms. Presentation copy printed on vellum.

E190 Invitation

The fifth Festival of Music and Poetry First concert will be given in the Music Room at Gregynog on Saturday, July 3 1937, at 8 p.m. A collection will be made in aid of the Orthopaedic Hospital, Oswestry

Single leaf. 90 × 115 mm. Baskerville type on card.

E191 Festival Programme

The fifth Festival of Music and Poetry at Gregynog 3–4th July 1937 [Device No 23] Contributions in aid of the Orthopaedic Hospital Oswestry will be accepted *Colophon*: Printed at the Gregynog Press July, 1937

[6] 18 pp. 285 × 192 mm. Perpetua type on grey silurian cover paper and Abbey Mills machine-made paper. The title is repeated on the cover, set in a pattern of rules and square dots. On the final text verso, 'Printed for private use at Gregynog, July 1937'. Colophon preceded by Device No. 24.

E192 Order of Service

The fifth Festival of Music and Poetry 3–4 July, 1937 [cross in red] An order of service to be used at Gregynog on Sunday 4th July 1937 [cross in red] *Colophon*: Gregynog July, 1937

[4] 14 [6] pp. 283 × 187 mm. Perpetua type on double grey cover paper and Abbey Mills machine-made paper. Text printed in red and black. On final text verso, 'Printed for private use at Gregynog, July 1937'. Colophon preceded by Device No. 23.

E193 Order of Service

In memory of Harriet Oliver died 24th June, 1937 An order of service to be used at Gregynog on Monday 5th July 1937 *Colophon*: Gregynog July, 1937

[4] 8 pp. 283 × 187 mm. Perpetua type on double grey cover paper and Abbey Mills machine-made paper. On final text verso, 'Printed for private use at Gregynog, July 1937'. Colophon preceded by Device No. 23.

E194 Order of Service

National Council of Music eighteenth annual conference 1–4 October 1937 [cross in red] An order of service to be used at Gregynog on Sunday October 3 1937 [cross in red] *Colophon*: Gregynog October, 1937

[2] 16 [2] pp. 280 × 187 mm. Perpetua type on double grey cover paper and Abbey Mills machine-made paper. On final text verso, 'Printed for private use at Gregynog, October 1937'. Colophon preceded by Device No. 23.

E195 Order of Service

League of Nations Union Welsh National Council Advisory Education Committee 20–22 November 1937 [cross in red] An order of service for use at Gregynog on Sunday November 21 1937 [cross in red] *Colophon*: Gregynog November, 1937

[4] 14 [2] pp. 280 × 187 mm. Perpetua type on double grey cover paper and Abbey Mills machine-made paper. Text printed in red and black. On final text verso, 'Printed for private use at Gregynog, November 1937'. Colophon preceded by Device No. 23.

E196 Order of Service
South Wales Girls' Camp 27–29 November 1937 An order of service to be used at Gregynog on Sunday November 28 1937 *Colophon*: Gregynog November, 1937
[2] 10 pp. 280×187 mm. Perpetua type on double grey cover paper and Abbey Mills machine-made paper. On final text verso, 'Printed for private use at Gregynog, November 1937'. Colophon preceded by Device No. 23.

E197 Conference Programme
The British Broadcasting Corporation Conference on Wales and Broadcasting Levavi [Device No. 12] oculos Gregynog December 4–6 1937 *Colophon*: Printed at the Gregynog Press December 1937
[12] pp. 210×140 mm. Perpetua type on double grey cover paper and Abbey Mills machine-made paper. Contains a list of delegates and the conference programme.

E198 Order of Service
The British Broadcasting Corporation Conference on Wales and Broadcasting 4–6 December 1937 [cross in red] An order of service to be used at Gregynog on Sunday December 5 1937 [cross in red] *Colophon*: Gregynog December, 1937
[4] 20 [4] pp. 283×189 mm. Perpetua type on double grey cover paper and Abbey Mills machine-made paper. Text printed in red and black. On penultimate text verso, 'Printed for private use at Gregynog, December 1937'. Colophon preceded by Device No. 23.

E199 Prospectus
The Gregynog Press announces The history of Saint Louis by John, Lord of Joinville Seneschal of Champagne translated from the French text edited by Natalis de Wailly by Joan Evans . . . The Gregynog Press MCMXXXVII
[4] pp. 343×240 mm. Poliphilus type on hand-made paper. Includes a description of the edition, a note about the book and a specimen page, printed in red, blue and black.

E200 Christmas Card
With greetings and best wishes for Christmas & the New Year from Gwendoline and Margaret Davies Gregynog 1937–1938
[6] pp. 280×220 mm. Perpetua type on blue cover paper and hand-made paper. Loosely inserted is a single leaf on which is printed a wood engraving of Virgin and child, measuring 166×100 mm, by Alison McKenzie.

E201 Christmas Card
On the morning of Christ's nativity by John Milton with a wood engraving by Alison McKenzie Gregynog MCMXXXVII *Colophon*: 250 copies of this poem have been printed at the Gregynog Press December, 1937
[4] 8 [4] pp. 275×203 mm. Perpetua type on cream cover paper and hand-made paper. Title repeated in blue on cover. Wood engraved frontispiece is of the Virgin and child. Text printed in blue and black. Loosely enclosed is a single leaf, measuring 270×193 mm, bearing the words 'With greetings and best wishes for Christmas and the New year from Gwendoline and Margaret Davies Gregynog 1937–1938'.

1938

E202 Order of Service

Social Reconstruction in South Wales ninth annual conference 12–14 February 1938 [cross in red] An order of service to be used at Gregynog on Sunday February 13 1938 [cross in red] *Colophon*: Gregynog February 1938

[4] 16 [4] pp. 284 × 187 mm. Perpetua type on double grey cover paper and Abbey Mills machine-made paper. Text printed in red and black. On penultimate text verso, 'Printed for private use at Gregynog, February 1938'. Colophon preceded by Device No. 23.

E203 League of Nations Union. Welsh Council. World Wireless Message

Dydd Ewyllys Da XVII Goodwill Day 18 Mai 1938 May 18 Neges flynyddol plant Cymru [wood engraving in red] Annual world wireless message trwy'r radio i'r byd of the youth of Wales Printed at the Gregynog Press

Single leaf. 280 × 365 mm. Perpetua type on hand-made paper. The wood engraving is a dragon which occupies two lines of type and divides Welsh and English titles. Text in Welsh and English, printed in two columns.

E204 Certificate

[Within a wood engraved chevron border] Eisteddfod Genedlaethol Urdd Gobaith Cymru [in red] Aberystwyth [in black] 9–11 Mehefin 1938 [wood engraving] Hyn sydd i arwyddo i o ennill clod ac anrhydedd yr eisteddfod yn am Gwasg Gregynog

Single leaf. 320 × 274 mm. Caslon type on hand-made paper. The border measures 170 × 185 mm. The wood engraving is a black outline shield containing a red dragon.

E205 Invitation

You are invited to the sixth Festival of Music and Poetry to be held at Gregynog 1–3 July 1938 Contributions in aid of the Orthopaedic Hospital and Derwen Cripples Training College, Oswestry will be accepted R.S.V.P. to the Secretary on the enclosed card before June 14.

[4] pp. 183 × 125 mm. Perpetua type on hand-made paper. The inner pages contain a list of events and artists.

E206 Invitation

The sixth Festival of Music and Poetry to be held at Gregynog 1938 Contributions in aid of the Orthopaedic Hospital and Derwen Cripples Training College, Oswestry will be accepted R.S.V.P. to the Secretary on the enclosed card before June 20.

[4] pp. 183 × 125 mm. Perpetua type on hand-made paper. On inside recto, 'You are invited to the Festival concert to be given in the Music Room – Gregynog on Friday July 1, at 8 p.m.'. This is followed by a list of artists.

E207 Invitation

The sixth Festival of Music and Poetry to be held at Gregynog 1938 Contributions in aid of the Orthopaedic Hospital and Derwen Cripples Training College, Oswestry will be accepted. R.S.V.P. to the Secretary on the enclosed card before June 20.

[4] pp. 183 × 125 mm. Perpetua type on hand-made paper. On inside recto, 'You are invited to the Festival concert to be given in the Music Room – Gregynog on Saturday July 2, at 8 p.m.'. This is followed by a list of artists.

E208 Festival Programme

The sixth Festival of Music and Poetry at Gregynog 1–3rd July 1938 [Device No. 12] Contributions in aid of the Orthopaedic Hospital Oswestry will be accepted *Colophon*: Printed at the Gregynog Press July, 1938

[9] 4–38 [4] pp. 285 × 189 mm. Perpetua type on blue-grey silurian cover paper and Abbey Mills machine-made paper. Cover title: [wood engraved numerals in blue, occupying five lines of type] VI [in black, set in a pattern of rules] The Gregynog Festival of Music & Poetry 1–3 July 1938. On penultimate text verso, 'Printed for private use at Gregynog, July 1938'. Colphon preceded by Device No. 24.

E209 Order of Service

The sixth Festival of Music and Poetry 1–3 July, 1938 [cross in red] An order of service to be used at Gregynog on Sunday 3rd July 1938 [cross in red] *Colophon*: Gregynog July, 1938

[4] 14 [2] pp. 284 × 190 mm. Perpetua type on double grey cover paper and Abbey Mills machine-made paper. Text printed in red and black. On final text verso, 'Printed for private use at Gregynog, July 1938'. Colophon preceded by Device No. 12.

E210 Order of Service

National Council of Music twentieth annual conference 23–26 September 1938 [cross in red] An order of service to be used at Gregynog on Sunday September 25 1938 [cross in red] *Colophon*: Gregynog September 1938

[4] 15 [5] pp. 285 × 190 mm. Perpetua type on double grey cover paper and Abbey Mills machine-made paper. Text printed in red and black. On penultimate text verso, 'Printed for private use at Gregynog, September 1938'. Colophon preceded by Device No. 23. 120 copies printed.

E211 Prospectus

The Lovers of Teruel[:] a drama in four acts in prose and verse by Juan Eugenio Hartzenbusch translated from the Spanish by Henry Thomas The Gregynog Press 1938

[4] pp. 245 × 175 mm. Bembo type on hand-made paper. Contains a note about the work, a specimen page and a description of the edition. 800 copies printed 13 October 1938.

E212 Prospectus

The Gregynog Press Forthcoming and recently published books The Press begs to announce that the following books are in active preparation: . . .

Single leaf. 130 × 182 mm. Perpetua type on hand-made paper. The verso lists *The praise and happinesse of the countrie-life*, *Shaw gives himself away* and *Parsifal*. The recto lists *The Star of Seville*, *The story of a red-deer*, *Caelica* and *The history of Saint Louis*. 800 copies printed 14 October 1938.

E213 Prospectus

The Gregynog Press announces the publication of The praise and happinesse of the countrie-life Written originally in Spanish by Don Antonio de Guevara Put into

English by H. Vaughan, Silurist Reprinted from the edition of 1651. Hand-set in 11-pt. Monotype 'Bell'. Introduction by Henry Thomas; seven head and tail-pieces, including title-page vignette, engraved on wood by Reynolds Stone . . . *Colophon*: The Gregynog Press Newtown, Montgomeryshire MCMXXXVIII
[4] pp. 182 × 123 mm. Bell type on hand-made paper. Includes a description of the edition, a note about the book and a specimen page.

E214 Order of Service
In memory of Lascelles Abercrombie M.A., Litt.D. 9 January 1881 – 27 October 1938 [cross in red] An order of service to be used at St. Martin-in-the-Fields on Monday, 12 December at 3.30 p.m. 1938 [cross in red] *Colophon*: Printed at the Gregynog Press December 1938
[4] 12 [4] pp. 283 × 190 mm. Perpetua type on double grey cover paper and Abbey Mills machine-made paper. Text printed in red and black. On penultimate text verso, 'Printed for private use'. Colophon preceded by Device No. 23.

E215 Christmas Card
Greetings and best wishes for Christmas and the New Year from Gwendoline and Margaret Davies Gregynog 1938–1939 *Colophon*: The Gregynog Press December 1938
[6] pp. 215 × 265 mm. Perpetua type on blue-grey silurian cover paper and hand-made paper. The covers enclose a loosely inserted leaf of hand-made paper, measuring 210 × 260 mm, on which is printed in blue and black lines from Chapter 28 of *The Book of Job* beginning, 'Where shall wisdom be found?' Colophon preceded by Device No. 28. 260 copies printed.

1939

E216 Order of Service
Social Reconstruction in South Wales tenth annual conference 28–30 January 1939 [cross in red] An order of service to be used at Gregynog on Sunday 29 January 1939 [cross in red] *Colophon*: Printed at the Gregynog Press January 1939
[4] 16 [4] pp. 281 × 188 mm. Perpetua type on double grey cover paper and Abbey Mills machine-made paper. Text printed in red and black. On penultimate text verso, 'Printed for private use at Gregynog, January 1939'. Colophon preceded by Device No. 23.

E217 Concert Programme
Gregynog Newtown Montgomeryshire The Passion according to Saint Matthew John Sebastian Bach 6th April 1939 The singing of The Passion will be preceded by a short service, and as the second part is to be broadcast, you are requested to be in your seat not later than 5.50 p.m.
[4] pp. 185 × 123 mm. Baskerville type on hand-made paper. Includes programme and list of artists.

E218 Order of Service
The Saint Matthew Passion John Sebastian Bach 1685–1750 [cross in red] An order of service to be used at Gregynog on Maundy Thursday 6 April 1939 [cross in red] *Colophon*: Printed at the Gregynog Press April, 1939

[4] 5 [7] pp. 283 × 187 mm. Perpetua type on double grey cover paper and Abbey Mills machine-made paper. Text printed in red and black. On penultimate text verso, 'Printed for private use at Gregynog, April 1939'. Colophon preceded by Device No. 23. 200 copies printed.

E219 League of Nations Union. Welsh Council. World Wireless Message
[In red] The world wireless message of the youth of Wales [in black] [Founded 1922] "Goodwill Day," May 18 The Welsh National Temple of Peace and Health Cardiff, Great Britain Date as postmark Dear Friend, . . .
Single leaf. 270 × 210 mm. Garamond type on hand-made paper. Consists of a letter of acknowledgment from Gwilym Davies, Vice-President. At the foot in red, '1939 Welsh League of Nations Union 1940' 600 copies printed.

E220 Certificate
[Within a wood engraved chevron border] Eisteddfod Genedlaethol Urdd Gobaith Cymru 1–3 Mehefin 1.9.3.9 Llan [wood engraving] elli Hyn sydd i arwyddo i o ennill clod ac anrhydedd yr eisteddfod yn am Gwasg Gregynog
Single leaf. 320 × 274 mm. Caslon type on hand-made paper. The border measures 170 × 185 mm. The wood engraving is a black outline shield containing a red dragon. 360 copies printed.

E221 Order of Service
Girls' Clubs in Wales 1–3 July 1939 [cross in red] An order of service to be used at Gregynog on Sunday 2 July 1939 [cross in red] *Colophon*: Printed at the Gregynog Press July, 1939
[4] 10 [2] pp. 280 × 188 mm. Perpetua type on double grey cover paper and Abbey Mills machine-made paper. Text printed in red and black. On final text verso, 'Printed for private use at Gregynog, July 1939'. Colophon preceded by Device No. 23. 100 copies printed.

E222 Prospectus
The Gregynog Press announces the publication of Shaw gives himself away[:] an autobiographical miscellany by George Bernard Shaw With a frontispiece engraved by John Farleigh This collection comprises three of the Prefaces, to which has been added a number of less familiar items here brought together from various sources. The text has been revised throughout by the author, who has written a preface to this edition. The Gregynog Press 1939
[4] pp. 250 × 170 mm. Baskerville type on green tinted hand-made paper. 2,000 copies printed.

1940

E223 Prospectus
The Gregynog Press announces the publication of Gweledigaetheu y bardd cwsc Ellis Wynne Visions of the sleeping bard Translated by T. Gwynn Jones With a frontispiece by Blair Hughes-Stanton . . . The Gregynog Press MCMXL
[4] pp. 275 × 208 mm. Bembo type on hand-made paper. Consists of a note about the work, two specimen pages, one in Welsh and one in English and a description of the edition in Welsh and English. 1,000 copies printed 28 June.

E224 Acknowledgment

E.E.H.J. 1919–1940

[4] pp. 185 × 123 mm. Bembo type on hand-made paper. On inside recto, 'Elsbeth Herbert Jones volunteered as an escort for the children being evacuated to Australia under the Government scheme last August. The "Rangitane", in which she was returning home, was sunk in the Pacific on November 27th, and she died from injuries received in the shelling. She was buried from the German raider on November 30th. Her family deeply appreciate the kindness shown to them in this time of sorrow, and send their warmest thanks to the many friends who have written expressing their sympathy.' Elsbeth was the only daughter of Dora Herbert Jones.

E225 Prospectus

Lyrics and unfinished poems by Lascelles Abercrombie 1881–1938 . . . The Gregynog Press MCMXL

[4] pp. 285 × 195 mm. Romulus and Cancelleresca Bastarda type on hand-made paper. Consists of a note about the work, a description of the edition and a specimen page. 1,000 copies printed.

E226 Concert Programme

A choral recital by the Gregynog Choir at the village hall Llandinam December 6th 1940 at 7.30 p.m.

[4] pp. 288 × 195 mm. Perpetua type on Abbey Mills machine-made paper.

E227 New Year Greeting

Resolutions when I come to be old – 1699 – . . . [17 resolutions from Jonathan Swift 1667–1745] from T. J. Gregynog Press. New Year's Day, 1941

Single leaf. 250 × 190 mm. Romulus type on hand-made paper.

1941

E228 Order of Service

In memory of Henry Walford Davies 6 September 1869 – 11 March 1941 [cross in red] An order of service to be used at Gregynog on Saturday 19 April 1941 [cross in red] *Colophon*: Printed at the Gregynog Press April, 1941

[4] 10 [2] pp. 282 × 190 mm. Perpetua type on double grey cover paper and Abbey Mills machine-made paper. Text printed in red and black. On final text verso, 'Printed for private use at Gregynog, April 1941'. Colophon preceded by Device No. 23.

Undated

E229 Poem

An ode to autumn by John Keats

Single leaf. 282 × 192 mm. Kennerley type on hand-made paper. Printed pre-1926.

E230 Speech

[Wood engraved initial I in black and sepia] t is not merely lecture-rooms and laboratories that are important. The place where those who are busy in the pursuit of different kinds of learning meet and observe each other are hardly less so. The union, the debating society, the talk with a fellow pilgrim on the steep and narrow way,

Opposite: Chrismas Card, 1928, printed in red and black (E43).

the friendship of these who are struggling to maintain a high level – these things, all of them, go to the making of a scholar. Words spoken by Lord Haldane in 1907, the year in which the Students' Union at Birkbeck College was established.
Single leaf. 220 × 305 mm. Perpetua type on hand-made paper. Printed 1930–1933.

E231 Psalm
Pourquoy son bruit Psalm II *Colophon*: Printed for private use at Gregynog
[8] pp. 215 × 285 mm. Bembo and Gregynog music type on hand-made paper. Consists of words and music printed on inner facing pages of two leaves of hand-made paper. The outer leaves consist of plain grey silurian paper covers. Colophon pasted on final text recto. Printed circa 1935.

E232 Festival Literature Insert
Accommodation can be reserved at the following hotels:— The Bear Hotel, Newtown Montgomeryshire The Buck Hotel, Caersws Montgomeryshire Early application, mentioning the Gregynog Festival, is advisable.
Single leaf. 92 × 109 mm. Baskerville type on card.

E233 Christmas Card
To wish you all that is best for Christmas and the New Year [wood engraving in blue] Jog on, jog on, the foot-path way, / And merrily hent the stile-a; / A merry heart goes all the day, / Your sad tires in a mile-a From Miss Margaret S. Davies Gregynog, Newtown
Single leaf. 173 × 125 mm. Perpetua type on Japanese vellum. The wood engraving incorporates the letters B.C. (Bovington Camp), a tent and sand dunes, within a circle. Probably printed 1930–1935.

E234 Envelope
[in red] [Device No. 7] If undelivered please return to the Gregynog Press Newtown, Montgomeryshire.
Envelope. 280 × 230 mm. Caslon type on cream cartridge paper. Printing appears in top left hand corner.

E235 Envelope
[in red] The Gregynog [Device No. 29] Press
Envelope. Kennerley type on cream wove paper. Printing appears in top left hand corner. Printed circa 1927.

WITH GREETINGS FOR CHRISTMAS AND BEST WISHES FOR THE NEW YEAR

Mrs. Edward Davies BRONEIRION

The Misses Davies GREGYNOG

MCMXXVIII

Press Devices

1

2 4

3

5
6 7 8
9

10
12 11 13
14

15 18 17
16 19 21
20 22

25 23 26
29
24 27 28

Further Reading

Blaker, Hugh. 'The Drawings and Engravings of R. A. Maynard', in *Studio.* Vol. XCIX, No. 443, February 1930. pp. 95–102.

Bowen, John Ewart. 'Memories of Gregynog', in *The Manchester Review.* Vol. VIII, Spring 1959. pp. 264–8.

Cave, Roderick. *The Private Press.* (London, Faber, 1971). 376pp.

Child, Harold. 'The Complete Book-Builder: a survey of the Gregynog Press', in *Studio.* Vol. XCIX, No. 445, April 1930. pp. 277–82.

Davies, Gwenllian M. 'The Gregynog Press: a Talk to the Manchester Society of Book Collectors, 16 October 1958', in *The Manchester Review.* Vol. VIII, Spring 1959. pp. 257–63.

Davies, Gwenllian M. 'Loyd Haberly: a Note on his Printing, plus a Checklist of Books from 1925 to 1960 Compiled by Loyd Haberly', in *The Manchester Review* Vol. IX, Autumn 1962. pp. 311–20.

Davies, J. Michael. *The Private Press at Gregynog.* (Leicester College of Art, 1959). 20pp.

Dunlap, Joseph R. *The Book That Never Was.* (New York, Oriole Editions, 1971). vi,86pp.

Fletcher, Ifan Kyrle. 'Two Books from the Gregynog Press', in *The Welsh Outlook.* Vol. XV, No. 1, January 1928. pp. 15–16.

Flower, Desmond. 'A Survey of Modern Binding', in *Signature.* No. 9, July 1938. pp. 19–32.

Franklin, Colin. *The Private Presses.* (London, Studio Vista, 1969). 240pp.

Gamble, William. 'The Gregynog Press', in *Penrose's Annual.* Vol. XXXI, 1929. pp. 129–31.

Haberly, Loyd. 'Loyd Haberly: Poet and Printer', in *The Printing Art.* Vol. I, No. 3, Autumn 1973, pp. 2–13.

Harrop, Dorothy A. 'George Fisher and the Gregynog Press', in *The Book Collector.* Vol. XIX, No. 4, Winter 1970. pp. 465–77.

Harrop, Dorothy A. 'The Gregynog Press', in *Encyclopedia of Library and Information Science.* Vol. X. (New York, Dekker, 1973). pp. 196–202.

Harrop, Dorothy A. 'The Gregynog Press', in *Gregynog,* edited by Glyn Tegai Hughes, Prys Morgan and J. Gareth Thomas. (Cardiff, University of Wales Press, 1977). pp. 95–118.

Herbert Jones, Dora. 'The Gregynog Press', in *Rural Industries*. Vol. XI, Summer 1928. pp. 3–4.

Herbert Jones, Dora. 'Sir John Morris Jones', in *The Welsh Outlook*. Vol. XVI, No. 5, May 1929. pp. 137–8.

Hutchins, Michael. *Printing at Gregynog: Aspects of a Great Private Press* (Welsh Arts Council, 1976). 40pp.

Irvine, A. L. 'Loyd Haberly and the Seven Acres Press', in *Book Handbook*. Vol. II, No. 3, September 1951. pp. 147–154.

Jenkins, David. 'Caniadau Gregynog', in *Y Traethodydd*. Vol. CXXVI, No. 538, January 1971. pp. 28–32.

Jenkins, David. 'The Gregynog Omar Khayyâm', in *The National Library of Wales Journal*. Vol. XVII, No. 1, Summer 1971. pp. 51–87.

Jenkins, David. 'Gwasg Gregynog', in *Llais Llyfrau*. Vol. VI, Winter 1966. pp. 1–5.

Jones, John Hugh. 'The Gregynog Press', in *Life and Letters Today*. March 1940. pp. 226–35.

Jones, John Hugh. 'The Gregynog Press', in *The Montgomeryshire Society Year Book, 1935–36*. pp. 42–47.

Jones, Thomas. 'Art in National Life: the Ubiquity of the Press in Wales: Makers of the Gregynog Books', in *The Western Mail*. 21 December 1925. p. 11.

Jones, Thomas. *The Gregynog Press: a Paper Read to the Double Crown Club on 7 April 1954*. (London, Oxford University Press, 1954). 40pp.

Jones, Thomas. 'The Gregynog Press', in *The Welsh Outlook*. Vol. XV, No. 8, August 1928. pp. 249–50.

Lewis, John. 'The Wood-Engravings of Blair Hughes-Stanton', in *Image*. No. 6, Spring 1951. pp. 26–44.

Maynard, Robert Ashwin. 'The Gregynog Press with a Private Type Foundry', in *The Monotype Recorder*. Vol. XXVIII, No. 232, September–October 1929. pp. 17–19.

Newdigate, Bernard. *The Art of the Book*. [Special Autumn Number of *Studio*] (London, Studio, 1938). 104pp.

Ransom, Will. *Private Presses and their Books*. (New York, Bowker, 1929). 493pp.

Ransom, Will. *Selective Check Lists of Press Books*. (New York, Duschnes, 1945). 420pp.

Symons, A. J. A. 'Modern English Bindings at the First Edition Club Reviewed', in *The Book Collector's Quarterly*. Vol. IV, No. 13, January–March 1934. pp. 81–6.

Tomkinson, G. S. *A Select Bibliography of the Principal Modern Presses*. (London, The First Edition Club, 1928). xxiv,238pp.

Williams, William. *Private Presses with Special Reference to Wales*. (London School of Printing, 1938). 12pp.

Index

References in bold type indicate illustrations.